Big Fleas have Little Fleas

Big Fleas have Little Fleas

or

Who's Who Among the Protozoa

by Robert Hegner

WITH A NEW INTRODUCTION BY

REGINALD D. MANWELL

Professor (Emeritus) of Zoology, and Research Associate in Syracuse University

Dover Publications, Inc., New York

Published in Canada by General Publishing Company, Ltd.,
30 Lesmill Road, Don Mills, Toronto, Ontario.
Published in the United Kingdom by Constable and Company, Ltd.,
10 Orange Street, London WC 2.

This Dover edition, first published in 1968, is an unabridged republication of the work originally published by The Williams & Wilkins Company in 1938. This edition contains a new Introduction by Reginald D. Manwell.

The publisher is grateful to the Directors of the Villanova College Library, Villanova, Pennsylvania, for supplying a copy of this work for the purpose of reproduction.

Standard Book Number: 486-22040-0
Library of Congress Catalog Card Number: 68-9783

Manufactured in the United States of America
Dover Publications, Inc.
180 Varick Street
New York, N. Y. 10014

INTRODUCTION TO THE DOVER EDITION

IT WILL be good news to all who are devotees of the biological sciences, but who also know that science has its lighter side, that this little book of Dr. Robert Hegner is at last being reprinted. To me it has always seemed a masterpiece of its kind, and there are all too few like it. Though based on a series of lectures given in 1937 at Cornell University, much of the material and many of the pictures were first used in a Sigma Xi lecture at Syracuse University, where they received an enthusiastic reception.

It was Dr. Hegner's purpose to make some of the important facts of protozoan biology intelligible to the interested layman, and he did it in a thoroughly entertaining and light-hearted way. However, biologists in specialties other than protozoology have always enjoyed the book too. Like other kinds of scientists, most professional biologists find that they are little more than laymen when they try to become informed in fields other than their own. Here the learning becomes fun.

Dr. Hegner makes us very aware of that zoological garden which each of us is likely to harbor within, a world which is completely beyond our ken without the microscope to reveal it, and of which we are ordinarily entirely unconscious. Some of the minute animals living there are simply boarders subsisting at our expense, doing us neither harm nor good. Others may benefit us to some small degree, perhaps by reducing slightly our dependence on outside sources of vitamins. Still others may reward us for the food and shelter we give by doing us to death. All, though uninvited and often unwelcome, are there because our negligent or unsanitary habits have made their entry possible. Knowledge of the way of life of these tiny organisms is therefore of great importance.

Dr. Hegner was in many respects a remarkable man. He had

an engaging smile and a cordial manner which those of us who were privileged to know him will never forget. He was an able writer, with numerous books and many papers to his credit. His *College Zoology*, first published in 1912 and reprinted and revised many times since, was one of the most successful college texts ever written. It introduced generations of students to zoology, and is still in use.

He was also a popular lecturer and a traveler to the far corners of the earth, where he diligently studied parasitic protozoa, particularly those of man and other primates. Some of these organisms are figured and described in this little book. They are the "lesser fleas," with their portraits depicted by a talented cartoonist.

Like many another biologist who achieved fame, Dr. Hegner was an able teacher. In his earlier years most of his students were undergraduates, but the last three decades of his life were spent at the School of Hygiene and Public Health of the Johns Hopkins University. Here he did research and taught graduate students, of whom I was fortunate enough to be one. Both Dr. and Mrs. Hegner were beloved by his students, for his home was often opened to them and his interest in them continued until his untimely death in 1943.

I am delighted that Dover Publications is reprinting this little book, and I am also happy to pay tribute to its author, to whom I owe much.

REGINALD D. MANWELL

Syracuse University
May, 1968

CONTENTS

Premonition 1

CHAPTER I

Out of the Everywhere into the Here 3

CHAPTER II

Parade of the Little Millions 30

CHAPTER III

And Little Fleas Have Lesser Fleas 57

CHAPTER IV

The Covered Wagon Marches On 83

CHAPTER V

Adam Had 'Em 117

CHAPTER VI

Monos de la Jungla 142

CHAPTER VII

Tell Me What You Eat 158

CHAPTER VIII

Oh, East is East and West is West 188

CHAPTER IX

Winged Guinea Pigs 215

CHAPTER X

Conquistadores 237

Appendix A. General Remarks about Protozoa 266

Appendix B. Glossary 270

Appendix C. Literature on Protozoa 275

Index 279

Big Fleas have Little Fleas

PREMONITION

BIG FLEAS Have LITTLE FLEAS
Upon Their Backs to bite 'em,
And LITTLE FLEAS Have LESSER FLEAS,
And so, AD INFINITUM.

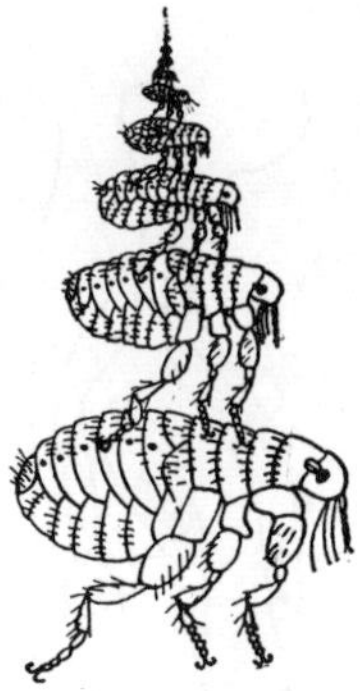

"Upon their backs to bite 'em"

FLEAS are pestiferous parasites and no respecters of persons. How doth the busy little flea? The question is asked and here's the answer:—Having travelled in certain less frequented parts of Mexico and other flea-infested countries, you can't fool us on fleas, we've had 'em. But the fleas also have 'em, perhaps not on their backs, but most certainly in their little bellies. What the fleas are to *Homo sapiens,* protozoa are to the fleas. Gnawing at their vitals are such fiends as the wily trypanosome, the ctenocephalus hemogregarine, the pulicid microsporidian, and the sinuous leptomonad. And do these denizens of the inner world lead a life of luxury and ease? The answer is no, most emphatically no. A few of them, at best, may for a time enjoy the privileges of the idle rich, but little realize that fate is lurking just around the corner, waiting patiently for them to stick out their necks.

But what are these funny little protozoa that everybody is talking about? Read about them here and marvel. If you would know how the other half live you must bore beneath the surface and visit these little and lesser "fleas" in their native haunts, among the muscle fibers, in the swift-flowing bloodstream, in the midst of the champing teeth, and within the peristaltic intestine.

But first you must know how the "first animals" (protozoa) came out of the everywhere into the here and must watch the little millions as they march in review before your microscopic eye. Then you may delve into the mysterious and often ignoble partnerships that nature imposes on the beasts and fowls, and "everything that creepeth upon the earth." You might now believe you are ready for man; but no, the monkeys must first be probed before you are properly prepared to deal with the lord of creation.

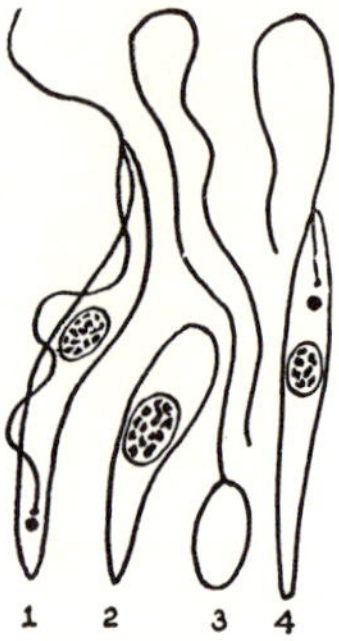

"In their little bellies." 1. Trypanosome. 2. Hemogregarine. 3. Microsporidian. 4. Leptomenad.

And now you think you have reached your goal. Alas, you have only completed the first of your Herculean labors; there are yet Augeias' Stables to be cleansed,—at least that is what it

FIG. 1. Alice might have seen something even more wonderful if she had looked through a microscope instead of through a looking glass.

may seem like to you before you have finished. What you eat determines the character of your hidden and unwelcome guests. Where you live decides what bugs shall occupy the department of the interior. Your training is not complete without a few struggles with winged guinea-pigs. But all your trials and tribulations will be forgot when you finally emerge as a full-fledged conquistador.

Chapter I

OUT OF THE EVERYWHERE INTO THE HERE

WHEN Alice stepped through the looking glass she encountered nothing more marvelous than the universe of protozoa ("first animals") that is revealed to us when we view through a microscope a bit of scum from the surface of a pond. Even a Lewis Carroll could not imagine such a variety of bizarre forms as these minute animals offer for our inspection. A world peopled with protozoa magnified 125,000 times, so that they would be equal in size to men and monkeys, horses and cattle, dogs and cats, and rats and mice, would be so extraordinary as to put to shame those imaginative artists who have attempted to visualize the possible inhabitants of Mars. Think of meeting a creature like *Amoeba proteus* of elephantine proportions slithering silently along Broadway, or picture the consternation of visitors to Coney Island if a ciliated behemoth were to arise suddenly from the briny deep. Under these conditions there would be no "great white way" and the seaside resorts would be forced into the hands of a receiver.

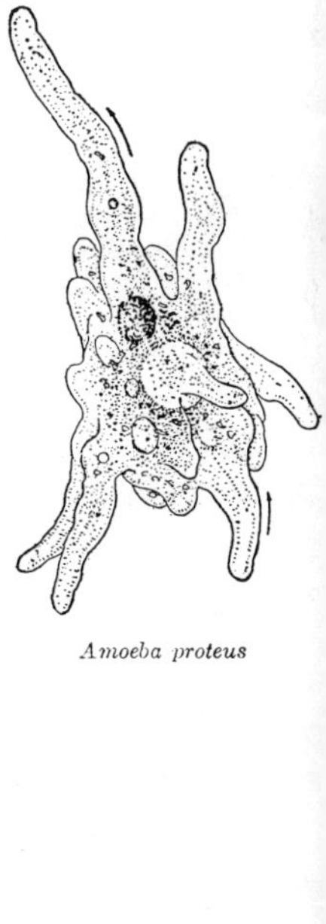

Amoeba proteus

PROTOZOA IN DRINKING WATER

So widespread and ubiquitous are protozoa on the earth it is impossible to escape from them no matter *how* carefully we regulate our lives. They are always present in the reservoirs from which we obtain our drinking water, sometimes in such numbers as to become a real nuisance. The green flagellate, *Euglena,* that is often so numerous as to impart its color to the fresh-water ponds in which it lives, is a comparatively inoffensive type. Some of the near relatives of *Euglena,* however, are not so innocuous. A species named *Uroglena* is one of the worst offenders, giving rise in drinking water to a fishy odor resembling that of cod liver oil. A branching species, *Dinobryon*, also produces a fishy odor. An odor resembling that of ripe cucumbers, and a taste both bitter and spicy, is furnished

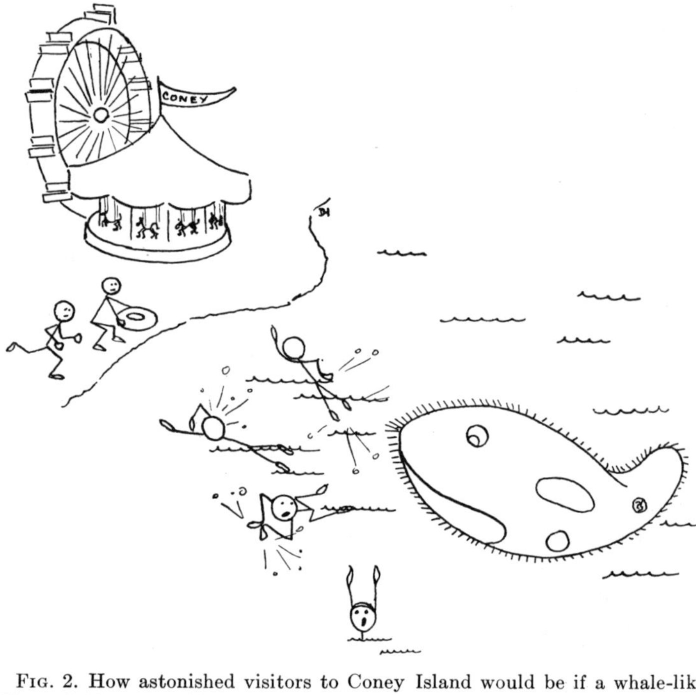

FIG. 2. How astonished visitors to Coney Island would be if a whale-like ciliate of this type suddenly appeared among the bathers.

by a species known as *Synura*. Chlorinated water drawn from a faucet may, after all, have some advantages over the water bright that comes from a spring-fed pond.

These tastes and odors are due to aromatic oils formed within the body and liberated when the animals disintegrate after death. An extremely minute quantity of this oil produces a perceptible odor; for example, one part of oil from *Synura* can be detected in 25,000,000 parts of water. The addition of copper sulphate in such small quantities as to be harmless to man and other animals, will destroy these protozoa and free the water from unpleasant odors and tastes.

PROTOZOA OF THE SOIL

We are all familiar with the expression "the good earth," but few of us are aware of the countless numbers of protozoa that live in this good earth with which we soil our hands when we work in the garden. If I had only known about them when

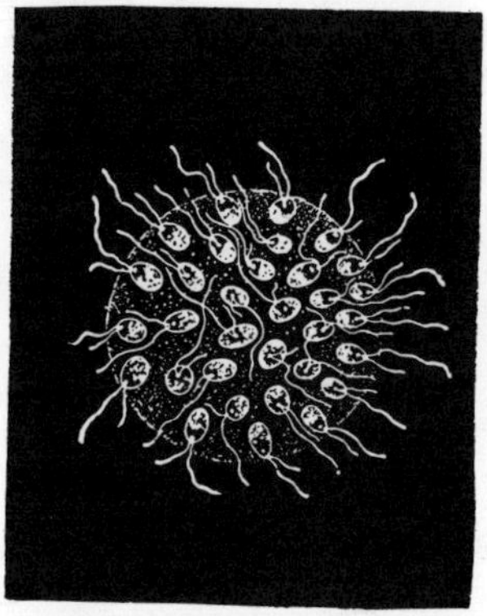

FIG. 3. Uroglena, a flagellate that sometimes gives a fishy odor to drinking water.

I was a boy perhaps I might have persuaded my father to let me play ball instead of forcing me to weed the onion bed. Two hundred and fifty different species of protozoa have been recorded from soils, some of them being represented by enormous numbers of individuals, and three of the four large classes into which the protozoa have been divided are represented. No soil, even the most barren, when properly examined has ever failed to disclose at least one species of protozoon. Ten samples of soil of different types were found by one inquisitive investigator to be inhabited by 1690 amebas, 7460 flagellates, and 15 ciliates per gram. If these were as large as meal worms or cheese skippers I am afraid garden clubs would not be as popular as they are nowadays.

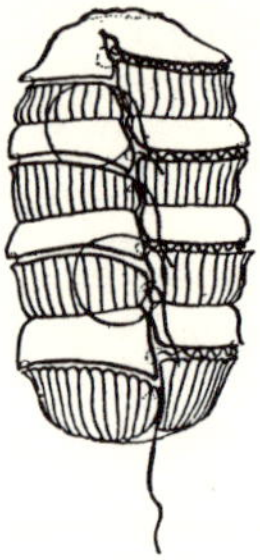

A marine protozoon

Soil protozoa may be of considerable practical importance, since a positive correlation exists between soil fertility and the number of protozoa present. In general, the greater the number of protozoa the more fertile is the soil. Since many soil protozoa feed on bacteria, a definite relation probably exists between these two types of living organisms, the exact nature of which has not yet been ascertained with certainty.

FIG. 4. Gadzooks! This water has both taste and smell

PROTOZOA IN THE SEA

We are accustomed to collect protozoa from ponds and streams, since most of us live inland, but the protozoa that float about in the sea easily outnumber the fresh-water species. Protozoa of the sea have more wide open spaces in which to flourish, inasmuch as 72 per cent of the earth's surface is covered by salt water. Furthermore, every inch of this space is occupied. Very few soil protozoa occur in the sub-soil, the maximum numbers being reached at a depth of about four inches, but the sea is inhabited from top to bottom. Not until we contemplate the depth of the ocean do we realize that the protozoa in the sea rival in numbers the stars in the Milky Way. The depth of the Atlantic Ocean averages about 12,000 feet; that of the Pacific reaches over 25,000 feet. Recently the protozoa were counted in samples of water taken from various depths in the Mediterranean Sea near Monaco where the bottom was located at 3000 feet. The average number of protozoa per quart of water from top to bottom was found to be about 4,000,000. If as many protozoa exist in the Atlantic Ocean as in the Mediterranean Sea a column of water four inches in diameter from top to bottom would contain 576,000,000,000

FIG. 5. How would you like to work in a garden with protozoa, like little gnomes, popping up at you on all sides?

individuals. This is a staggering figure even though the New Deal has taught us to think in terms of billions. If these protozoa could only be transmuted into dollars we could pay our national debt and vote a bonus of 4,000 dollars to every man, woman and child in the United States.

PHYSIOLOGY AND REPRODUCTION OF PROTOZOA

Even more amazing than their numbers and fantastic shapes are the activities of these microorganisms. Protozoa are said to be unicellular animals, each individual corresponding in this respect to one of the billions of cells that constitute the tissues in the bodies of man and other higher animals. From the structural standpoint this is correct, since a cell is a particle of living matter, protoplasm, containing a nucleus and, with few exceptions, each protozoon is a comparatively simple-looking particle of protoplasm containing a nucleus. Physiologically, however, unicellular protozoa are comparable to multicellular animals, since they are capable of carrying on with their single cell all of the processes necessary for the maintenance of the individual and of the race. They find food, capture it, ingest it, digest it, egest the indigestible matter and assimilate that which is digested. They carry on processes of respiration, secretion, and excretion. They react in definite ways to changes in their environment enabling them to find a favorable habitat, and to protect themselves from their brothers and sisters, their cousins and their aunts, and from other species of animals. And they reproduce their kind in enormous numbers.

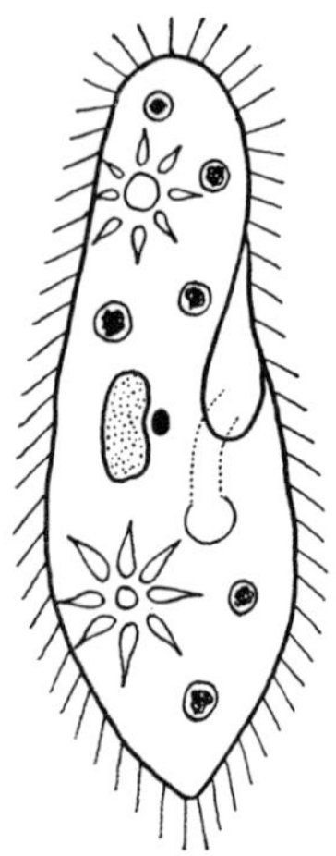

Paramecium

For example, the common slipper animalcule *Paramecium* grows and reproduces comparatively slowly; nevertheless, in one month a single specimen might become the progenitor of 268,000,000 offspring and soon the mass of paramecia would become as great as the known universe if all of these offspring were able to live and reproduce without interference. This type of mass production had been going on for millions of years before the automobile was invented. How envious Henry Ford and General Motors would be, if they but knew how many free-wheeling ciliates a *Paramecium* can turn out in a given length of time! However, there is an everpresent business depression in nature that limits the food supply with no beneficent government to provide a full dinner pail, hence only a few individuals succeed in the struggle for existence,—just enough protozoa survive to keep the race from dying out and to provide biology teachers with material for their classes, and investigators with a supply of organisms about which they can write learned papers for scientific periodicals. Such an egotistical animal as man naturally considers himself the dominant type of life, but if numbers of species and individuals and ability to live under diverse circumstances are accepted as criteria of dominance, paramecia and other protozoa are manifestly more successful than are human beings. I feel so convinced of this that I would not hesitate to nominate one of these humble creatures as President of the World.

NUMBERS OF SPECIES OF PROTOZOA

It seems really remarkable that an animal such as ameba that is only about 1/100 of an inch in diameter, and that looks

FIG. 6. A high-powered struggle for existence is continually in action among protozoa and only one in millions survives.

like a minute particle of animated jelly, can maintain itself in the struggle for existence in competition with the hundreds of thousands of species of larger animals that inhabit the earth; yet, as every one knows who has studied zoology, *Amoeba proteus* is one of the most widely distributed and most abundant of all known animals, and its fitness to survive cannot be challenged. But *Amoeba proteus* is only one of about 15,000 different species of protozoa known to science. This, moreover, does not furnish an adequate idea of the number of different species that really exist today. Swarms of parasitic species that live on or within the bodies of other animals have evolved during the ages from the free-living protozoa. How many different species of these parasitic protozoa there are can hardly be estimated. We know that human beings may be inhabited

FIG. 7. Because they are dominant in number of species and individuals a protozoon is hereby nominated President of the World.

by at least twenty-five different species; that domesticated animals may be parasitized by as many or more; and that practically every wild animal that has ever been examined properly supports one or more species. Furthermore, the species that live in one host usually differ from those that live in other species of hosts; there must therefore be at least as many species of parasitic protozoa as there are other species of animals available to parasitize. Protozoologists, like the Walrus and the Carpenter, are tempted to weep when they contemplate such quantities of "bugs" pleading to be discovered.

PROTOZOA IN THE EARTH'S CRUST

Although Protozoology is one of the youngest of the biological sciences and protozoa have been known by this name for a comparatively short period, the effects of certain species of protozoa on human beings are recorded in ancient chronicles. Malaria, for example, which is due to protozoa that live in the red blood corpuscles, was well known to the Greeks and Romans 2000 years ago, although the specific organisms responsible for the disease were not discovered until 1880, and their method of transmission through the agency of mosquitoes was not determined until 1898. Certain fossil protozoa, the nummulites, were known to the ancient Egyptians who built their pyramids of limestone that consisted largely of the shells

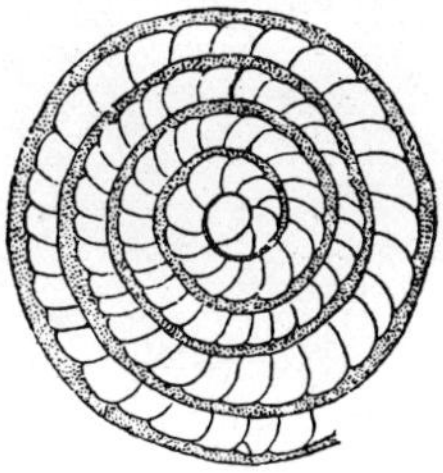

Nummulite shell

of this type. According to the geographer Strabo (first century B.C.) a tradition existed to the effect that these nummulites were petrified lentils discarded by the pyramid builders. Nummulitic limestones now extend across Southern Europe, Egypt, Asia Minor and the Himalayas. One's credulity is sorely stretched when he is asked to believe that a tiny mite of protoplasm can build up such an elaborate apartment house as this. The whorls are coiled in a spiral and each whorl completely covers the preceding one so that only the last whorl is visible externally, just as the walls of a building conceal the rooms within. The windows and doors of the shell are represented by minute pores which allow communication among the chambers and with the outside. Evidently these nummulites allowed their architectural ambitions to overcome their ability to maintain such elaborate structures, since, of the vast multitudes that once inhabited the primordial seas, only one or two rare forms remain among the living.

Even today certain protozoa are busily engaged in building up the earth's crust. A common type are the globigerinas. These homeless orphans float about in the sea in enormous numbers being carried wherever waves and currents may take them. Most of their lives are spent in secreting a shell of carbonate of lime in which to live. When the first globe-like chamber becomes too small, another larger one is constructed

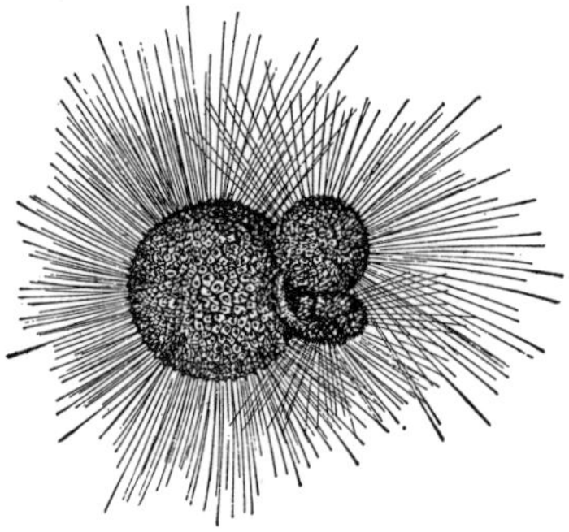

FIG. 8. Globigerina with its hundreds of radiating spines must seem like a marine porcupine to its aquatic neighbors.

and soon there is built up a group of sieve-like apartments, each larger than the last; these are protected with long, sharp spines. Globigerinas do not, like the chambered nautilus, leave the "past year's dwelling for the new," but completely fill all of their many-chambered residence. The labor involved in extracting such an enormous amount of calcium carbonate from the water must be tremendous.

In course of time the globigerinas die, and their shells sink to the bottom of the ocean, where they accumulate in such enormous numbers as to form a thick layer that is known as Globigerina Ooze. The greater part of the bottom of the Atlantic Ocean is covered with this ooze—an area of about

20,000,000 square miles. As the deposits increase, the ooze becomes compact and finally solidifies in the form of limestone.

In the Pacific and Indian oceans an area of about 2,290,000 square miles of sea bottom is covered with Radiolarian Ooze. This is an accumulation of shells built up as are those of globigerina except that they consist of silicon instead of calcium carbonate. Some idea of how much labor is required of these minute specks of living jelly to build up their shells is indicated by the fact that silicon exists in solution in sea water in only about $1\frac{1}{2}$ parts in 100,000, hence a ton or more of sea water is necessary to provide a single ounce of skeleton.

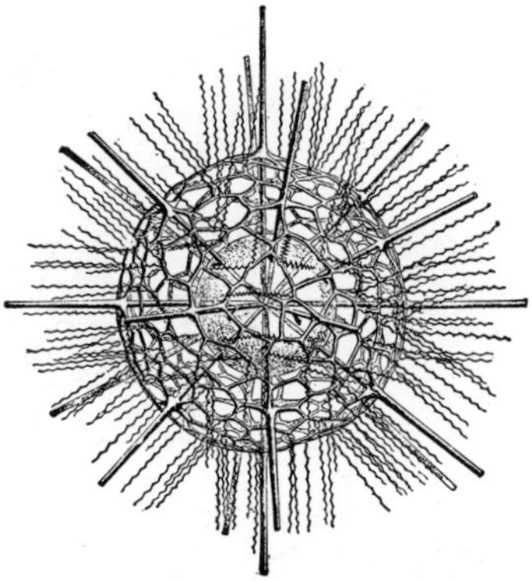

FIG. 9. This beautiful skeleton of silicon was built by a radiolarian. One would hardly expect to find such a delicate organism in the raging seas.

LEEUWENHOEK AND THE DISCOVERY OF PROTOZOA

One of the most exciting periods in the history of biology was that which witnessed the development of the microscope, since this instrument brought to light a microcosmos entirely unsuspected. How thrilled the naturalists of that day must have been when they examined all sorts of flotsam and jetsam and found it to be alive with myriads of beautiful little creatures moving about as in a Lilliputian metropolis! They must have felt like real big-game hunters as they searched the micro-jungle for these strange types of wild animals. To them the newly discovered protozoa appeared more wonderful even than they seem to us, since their bodies were supposed to be provided with systems of organs similar to those of higher animals. "How marvelous," said Leeuwenhoek, "must be the visceral apparatus shut up in such animalcula!" No wonder he was overwhelmed by what he saw when he first looked through his primitive microscope. Even today the study of

FIG. 10. Viewing a chaotic collection of protozoa through a microscope for the first time is enough to overwhelm even a Dutchman.

protozoa is not without its thrills and the pioneer spirit of the early investigators is easily captured by anyone who studies for the first time the protozoan inhabitants of some new species of host; in many cases these protozoa have never been seen previously by anyone else, and have waited impatiently for centuries for scientists to reveal them to the world.

The Dutch microscopist Anthony van Leeuwenhoek is justly

known as the Father of Protozoology. He was born in Delft in 1632 and at the age of twenty-two started in business as a draper. Microscopes interested him and the grinding of lenses became one of his hobbies. With his lenses he examined practically everything he could think of, and, since he possessed an expansive imagination, naturally made many interesting discoveries. In 1674, at the age of forty-two, he discovered free-living protozoa in a fresh-water pond, including what must have been *Euglena viridis* and several species of well-known ciliates. The same year he described parasitic protozoa for the first time in material obtained from the gall bladder of a rabbit; his "oval corpuscles" were no doubt the oocysts of the coccidium of the rabbit, *Eimeria stiedae.* In 1681 he discovered human protozoan parasites in his own excrement and described what one can easily recognize as the intestinal flagellate *Giardia lamblia.*

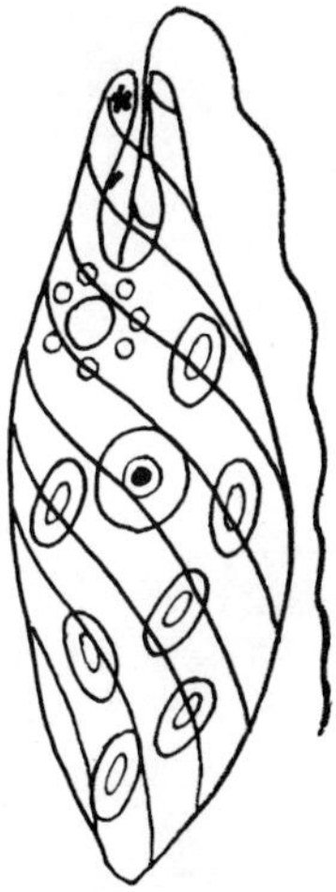

Euglena

These and many other discoveries were recorded mostly in the form of letters to the Royal Society of London. These letters were written in Old Dutch, hence the unfortunate protozoologist of the present day who wishes to read them must first learn this formidable language. Leeuwenhoek's success was due to his keen powers of observation and his wonderful skill in grinding lenses. He left behind him 247 complete microscopes and 172 mounted lenses. The best of these magnified from 270 to 300 diameters.

Following Leeuwenhoek came a long series of investigators who gradually added to the knowledge of protozoa until the group was fairly well delimited. Several of these investigators were among those who will be known for all time as major contributors to the science of Biology.

DUJARDIN AND THE DISCOVERY OF SARCODE

Dujardin, for example, recognized that the living contents of protozoa is of greater importance than the cell wall and to this substance in 1835 he gave the name *sarcode.* The following quotation taken from his paper entitled "On the Lower Organisms," and translated into English, appears to contain the first reference to this name. At this time Ehrenberg and others were insisting that protozoa were filled with viscera including esophagus, stomachs, intestines and anus. To be sure the anal aperture of *Paramecium* was soon found to be only temporary, but as a student of mine once remarked, a temporary anal aperture is better than none at all.

Dujardin writes as follows: "The difficulty in determining the nature of these stomachs or globules seemed to me for a long time insurmountable—and since I make use of instruments as perfect as obtainable and have had long experience in this type of observation, I blamed myself for not being able to distinguish the intestine to which all these stomachs should be attached, nor the anal and buccal orifices of this intestine. The high respect due to the works of Ehrenberg did not permit me to question structures used by him as the foundation of his classification. As a matter of fact, I knew other observers skillful in the use of the microscope, who were no more fortunate than myself; nevertheless I might have given up hope and abandoned this research as altogether out of proportion to my vision, if I had not fortunately found the solution of the problem in the discovery of the properties of Sarcode."

"I propose to call by the name Sarcoda that which has been called by other workers living jelly; that viscid translucent substance, insoluble in water, contracting into globular masses, sticking to dissection needles and forming strings like mucus, found in all lower animals where it forms a matrix for the other elements of structure."

As not infrequently happens in science the name coined by the original discoverer was superseded by one suggested later. If the laws of priority in the field of zoology operated with respect to structure and function as it does with respect to genera and species, living substance would now be known as sarcode, not protoplasm.

PROTOZOA AND SPONTANEOUS GENERATION

Another period in the history of biology during which protozoa played an important rôle was that which witnessed the defeat of the proponents of spontaneous generation.

Out of the Everywhere into the Here

Previous to the end of the seventeenth century the doctrine of the Spontaneous Generation of animals was accepted by both scientists and philosophers. According to this doctrine many species of animals were supposed to arise spontaneously from non-living matter. For example, insects were believed to originate from dew, frogs and toads from the muddy bottom of ponds under the influence of the sun, skippers from cheese and fly maggots from flesh. Even today many uneducated people believe that mosquitoes are generated by stagnant water and that horse hairs, when placed in water, change into living

FIG. 11. The "groundhogs" who believe in spontaneous generation saw their shadows when Redi and Pasteur entered the lists, and lost no time in diving back into their holes never again to emerge into the light of day.

"snakes." The classical experiment of the Italian, Redi, in 1680 exploded this theory as regards the larger types of animals. Redi placed some meat in a jar and covered the entrance with gauze. Flies were attracted to it but could not reach it and hence their eggs were laid on the gauze and no maggots appeared in the decaying meat.

The advocates of spontaneous generation were thus chased into their holes by Redi, but when Leeuwenhoek discovered protozoa, these misguided reactionaries decided that spring had come again and at once emerged from their winter quarters in large numbers. To be sure protozoa were soon observed to multiply by fission and not to arise *de novo* from decaying vegetation, but objections of many sorts had to be overcome before every one was finally convinced that all life arises from preëxisting life. The discovery of bacteria complicated the situation, but the sceptics were eventually silenced by Pasteur. He demonstrated that the microörganisms that appeared in sterilized infusions came from the "dust" in the air, and that no life arose in these infusions when they were properly protected.

If protozoa do not arise from non-living matter, it is natural to inquire how the world became populated with them, since both geologists and astronomers tell us that at one time long ago life could not have existed on the earth. The doctrine of Special Creation, that is, that each species of animal was specially created, is sufficiently refuted by the facts of organic evolution to require no discussion. Protozoa must, therefore, have originated on the earth from non-living matter or been brought to the earth from some other part of our universe. The latter, known as the Cosmozoa Theory, is so improbable as to be hardly worthy of consideration; even if it were true, it does not explain the origin of protozoa but only how protozoa reached the earth.

A number of theories have been proposed to account for the origin of protozoa on the earth, especially during the period when the earth was cooling down from its original incandescent condition. Even now protozoa might conceivably arise *de novo* from non-living matter if the various elements contained in protoplasm were to chance to unite in the proper quantities and in the proper relations to one another, but actually we have no real evidence and must conclude that we do not know when or how protozoa originated and probably will not know for many generations to come. The protozoa that now exist on the earth differ sufficiently from one another so that they can be recognized as distinct species.

Zoologists do not agree as to the exact meaning of the term species. One authority gives the following definition: "A species may be defined as a group of interbreeding individuals which, while they may differ markedly among themselves, yet resemble each other more closely than they do those of any other group; the characters that distinguish the group being considerable, not obliterated by intermediate forms, and inherited from generation to generation." A shorter definition is, "a species is a group of organisms differing from one another only in age, seasonal, sexual or individual characteristics."

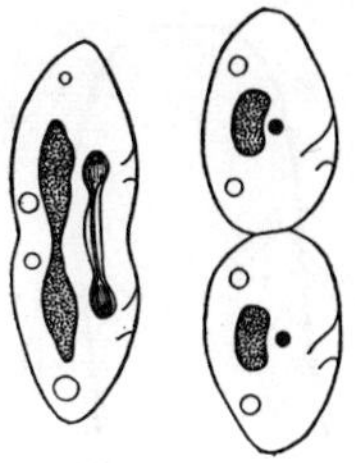
Binary fission of Paramecium

IMMORTALITY OF PROTOZOA

Not long after the defeat of spontaneous generation, the belief in the immortality of the protozoa became prevalent. Apparently the only animals that are potentially immortal are the protozoa. As every zoologist knows, in many cases the substance of the parent protozoon lives on in the bodies of her daughters and no natural death occurs. For example, the common paramecium reproduces by binary division and apparently in no other way. When it reaches a certain size the chromosomes in the nucleus divide each into two with great exactness. Two nuclei are formed each with a complete set of daughter chromosomes. Then the body divides into two. During division the mother of course loses her individuality, but she lives on in the bodies of her offspring. This has been going on since paramecium first appeared on the earth and will probably continue until the world comes to an end.

Not all protozoa are potentially immortal, since as we shall see later, the intervention of sex and the development of division of labor among cells introduced phenomena that eventually lead to the natural death of the body, although, even in higher animals, the germ plasm is continuous from generation to generation.

Modern children demand that they be allowed to live their own lives. The precocious offspring of the protozoa, however, go a bit further, since each daughter not only lives her own life but also one-half of her mother's, one-fourth of her grandmother's, one-eighth of her great grandmother's and so *ad infinitum*. There must be little satisfaction in maintaining one's individuality for one short day only to be cut in two by a cell wall so that the Freshman class in biology may be provided with plenty of busy work.

Immortality of Paramecium

THE STUDY OF ORGANIC EVOLUTION WITH PROTOZOA

Darwin did not make extensive use of protozoa in his attempt to account for the origin of species by the theory of natural

FIG. 12. To the biologist only the protozoa are immortal, but who wants to be immortal if he loses thereby his individuality?

selection, but recently several types of evidence have been secured that have an important bearing on organic evolution. Every biologist would like to see organic evolution taking place but few have witnessed such a phenomenon; among those who have are several investigators who have studied fresh-water sarcodina. I had the good fortune to take part in these researches, using the beautiful shelled type *Arcella* as a protozoan guinea-pig. The name *Arcella* means "a little box." This little box is to arcella what a house is to us and what the shell is to an oyster. It is something to retire into to escape bad weather or enemies.

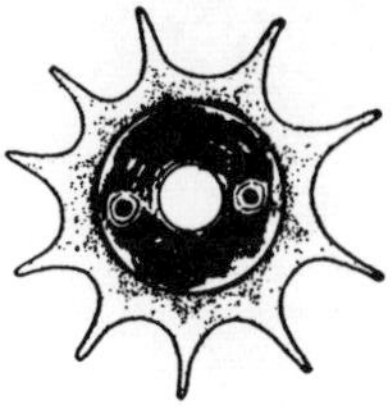

Arcella dentata with 10 teeth, 2 nuclei and dark-colored cytoplasm.

The shell of arcella is a very beautiful light-brown structure, which, when looked down upon through a microscope, reminds one somewhat of a doughnut. The upper surface is shaped like a dome. In the lower surface is a circular opening, the mouth of the shell, out of which the false feet are thrust in search of food or for crawling about among the weeds in the pond where it lives.

Reproduction in arcella is similar to that in *Amoeba proteus* but is made more complex by the restriction of the shell. When binary division occurs most of the protoplasm within the shell flows out and organizes itself into a mass shaped like that of the shell. It then secretes a new shell at the surface and so we find parent and daughter shells lying mouth to mouth. Soon one-half of the protoplasm flows back into the parent shell and mother and daughter separate.

And now I shall attempt by means of diagrams to explain even to those readers who are not biologists how several distinct types of arcellas were encouraged to evolve from a single ancestor. Although incapable of mastication, the particular species of *Arcella* worked with is well supplied with teeth; these are located around the edge of the shell and hence the species is known as *Arcella dentata*. Variations in the number of teeth were noted among wild arcellas, hence this characteristic was chosen as a basis for our experiments. A single specimen was allowed to divide in a culture dish and its two daughters were isolated as soon as the division was completed. These daughters were likewise allowed to divide and their progeny isolated. Thus a number of what geneticists call pure lines, or clones, were obtained from a single arcella. The number of teeth was counted on the descendants of these arcellas, and after many generations had been produced, all were discarded except two pure lines, one with the lowest average number of teeth and the other with the highest average number of teeth. Specimens were then selected generation after generation from each of these pure lines, on the one hand with the lowest and on the other hand with the highest average number of teeth.

After a period of 64 days and 22 generations one pure line contained specimens with an average tooth number of about 13 and the other contained specimens with an average tooth number of about 9. These two pure lines were then allowed to multiply by binary division without any selection whatever. After a period of 35 days the average number of teeth in the two pure lines was practically the same as when selection was discontinued. From this we concluded that by selection we

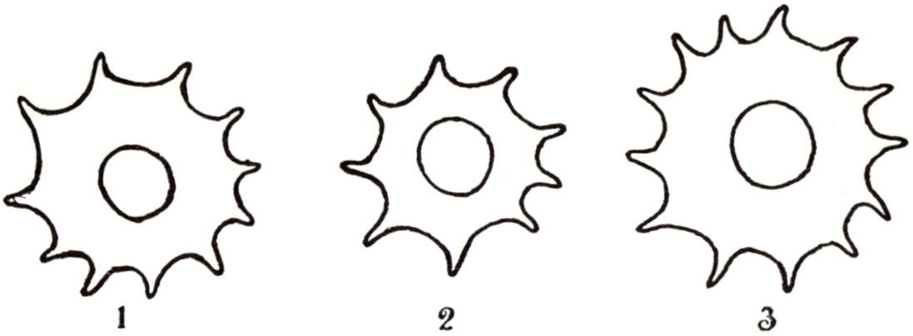

FIG. 13. Shells of *Arcella dentata*. 1. Original parent. 2. Example from smaller pure line. 3. Example from larger pure line.

had obtained from a single parent two types of offspring that differed in the number of teeth and that this difference was inherited. This means that we had been selecting variations that were inherited by the offspring. Biologists now believe that organic evolution has taken place in nature, at least in part, by the natural selection of small heritable variations. If this is correct, then we have observed in the laboratory the evolution of two distinct types of arcellas from a single type.

Our results were obtained in a comparatively short time largely because multiplication in arcella takes place so rapidly, one generation following another at intervals of only about three days, whereas when working with higher animals one must often wait for weeks, months or even years before the young reach the age of reproduction.

EVOLUTION OF PARASITIC PROTOZOA

The opportunities offered by parasitic protozoa for the study of evolution are many and varied. Of special interest is the

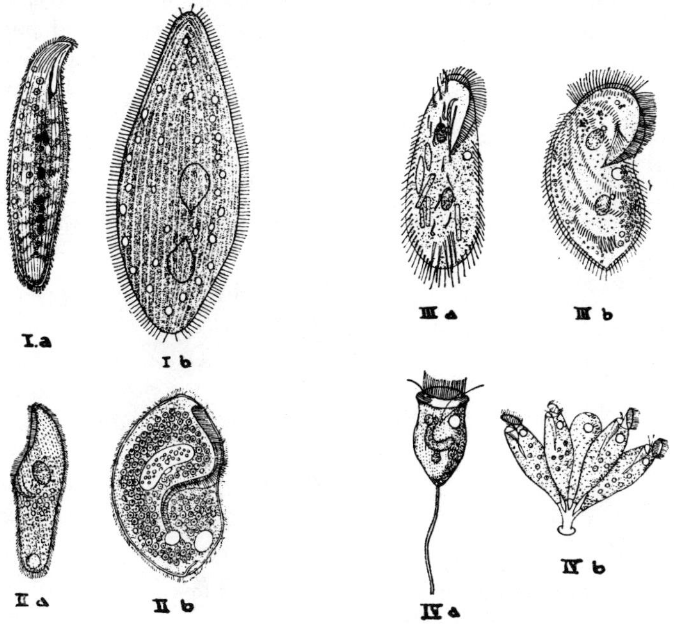

FIG. 14. These ciliates are arranged in pairs. Each pair consists of two species belonging to the same family, at the left a free-living species and at the right a parasitic species. The parasitic species probably evolved from similar free-living species.

evolution of the parasitic habit. Ciliates, for example, are thought of by most zoologists as free-living organisms, but the orders, and even some of the families, within the class Infusoria contain both free-living and parasitic species. It seems evident that the parasitic species have evolved within each order or family from free-living species. Thus the parasitic habit must have evolved independently within each group. This type of evolution has therefore taken place many times and must be comparatively easily accomplished.

For example, the accompanying diagrams include a free-living and a parasitic species of each of several orders. Thus in the order Holotrichida, which contains ciliates with cilia of approximately equal length on the surface of the body, are located free-living species such as *Loxodes rostrum* and parasitic species like *Amphileptus branchiarum* that lives on the gills of tadpoles. These two species even belong to the same

FIG. 15. Parasitic protozoa are proud of their free-living ancestry

family, the Tracheliidae. Among the Heterotrichida, which bear longer cilia around the mouth opening, are such species as *Blepharisma lateritia* that lives in fresh water and *Nyctotherus ovalis* that inhabits the intestine of cockroaches. These belong to the family Plagiotomidae.

A third order, the Peritrichida, which possess a spiral zone of cilia around the mouth opening, includes such fresh-water species as *Vorticella alba,* and also ectoparasites that cling to the surface of certain aquatic animals and are known as *Opercularia stenostoma.* These are numbers of the family

Vorticellidae. *Kerona pediculus,* of the order Hypotrichida, is a louse-like species that creeps about as an ectoparasite on aquatic animals. This species and its free-living relatives, such as *Oxytricha fallax,* are much-flattened protozoa with cilia fused together to form leg-like cirri. They belong to the family Oxytrichidae.

Apparently there is no question as to the evolution of the parasitic species in each of these orders and families from free-living species that belonged to the same orders and families.

Protozoa also occupy an important place in the history of our knowledge of genetics. Nowadays we hear a great deal about that habitual drunkard of the insect world, the fruit fly *Drosophila,* and especially about the structure of its salivary glands, but paramecia and certain other protozoa were revealing to us secrets of heredity long before it was suspected that *Drosophila* and its salivary glands were worthy of study.

TYPES OF ASSOCIATION BETWEEN PROTOZOA AND OTHER ANIMALS

Protozoa furnish particularly favorable material for the study of the types of association known as commensalism, symbiosis and parasitism. The relation between the protozoon, which is usually considered to be the parasite, and the host with which it is associated, ranges from a facultative, incidental type of companionship to an obligatory partnership that may be mutually beneficial or may result in serious injury to the host and even death. Thus commensals live together amicably; one member of the association is benefited and the other is neither harmed nor benefited. One student on the basis of this description suggested companionate marriage as an example of commensalism. Animals that are mutually beneficial are said to be symbiotic. A true parasite, strictly speaking injures its host.

Some years ago I took the trouble to compare the structure and activities of free-living protozoa with those of parasitic protozoa, and reached the conclusion that no fundamental differences exist between these two types. The most obvious difference is that of habitat. Parasitic protozoa live in or upon the bodies of other organisms. The study of parasitic protozoa has not only been of great biological significance but has also resulted in some of the most conspicuous successes in the field of preventive medicine, such as the control of malaria, amebic dysentery, African sleeping sickness and kala-azar. Since these will be discussed later they need not be described at this time.

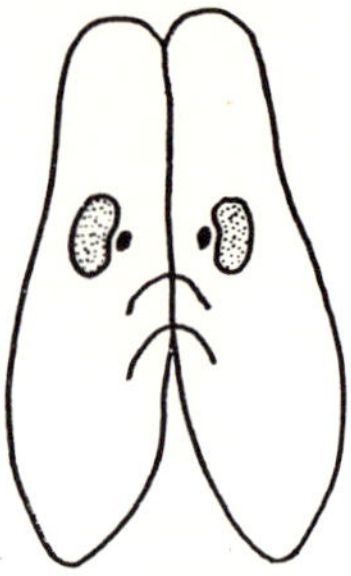

Conjugation of Paramecium

SEXUAL PHENOMENA IN PROTOZOA

Everyone who lives beyond the first decade is more or less interested in sex and as he grows older becomes, in course of time, aware of the approach of senility. Biologists have recognized the importance of these subjects both from the practical and scientific viewpoints, and, strange as it may seem, have found the protozoa to be excellent material for studying these rather intimately related phenomena. So universal is sex among animals that already among the protozoa sexual differences between male and female cells can be distinguished. Of particular interest is the question whether the union of the two sexes is necessary for continued existence and brings about rejuvenescence.

Paramecia has been used for the study of this problem more than any other species. These ciliates may be said to be ambisextrous, since when conjugation occurs, each fertilizes the other. At one time it was supposed that this temporary union of two paramecia during which nuclei are exchanged was necessary to prevent the senescence and death of these organisms. Experiments have proved, however, that if kept in a proper environment, paramecia may retain their vigor year after year and apparently indefinitely without the intervention of sexual union. Thus fertilization is not necessary for the continued existence of the race although when living under adverse conditions the paramecia seem to be invigorated by the process.

What then is the significance of the union of sexual cells? A study of the offspring of the paramecia after fertilization gives us what appears to be the correct answer. The chief object of this phenomenon is the reorganization of the nuclear material. The results of this reorganization seem quite definite. If a single *Paramecium* is selected and allowed to reproduce by binary fission, its descendents which form a clone are similar in constitution. If two individuals belonging to one clone conjugate and the exconjugants are allowed to reproduce by binary fission and their descendants studied carefully, they may be found to differ in various respects. For example, one exconjugant may give rise to descendants that divide more rapidly than the rate characteristic of the clone, and other exconjugants may give rise to descendants that divide less rapidly. Other diversities, both physiological and morphological, may also appear. Conjugation leading to nuclear reorganization thus results in diversities in the exconjugants that are inherited by their offspring.

Sexual processes in protozoa therefore lead to changes in the germ plasm as a result of the crossing of two different individuals, a phenomenon that involves a redistribution of the genes of inheritance responsible for evolution. This conclusion is clearly indicated as a result of investigations on these unicellular organisms. No wonder higher animals have become differentiated by nature into males and females and have been so liberally endowed with sexual instincts.

BEHAVIOR OF PROTOZOA

Most people are surprised when they learn that anyone is interested in why protozoa behave like protozoa. Nevertheless it is a fact that many scientists have spent a large part of their lives engaged in studying the reactions of these lowly organisms to the changes in their environment. The behavior of a protozoon represents the "animal mind" in its most primitive form. By behavior we mean the sum total of the animal's reactions to stimuli. These reactions are responses to changes in the environment and are due to a fundamental property of protoplasm, called irritability. In higher animals certain cells are specialized for the sole purpose of performing nervous functions but in protozoa the organism acts as a whole.

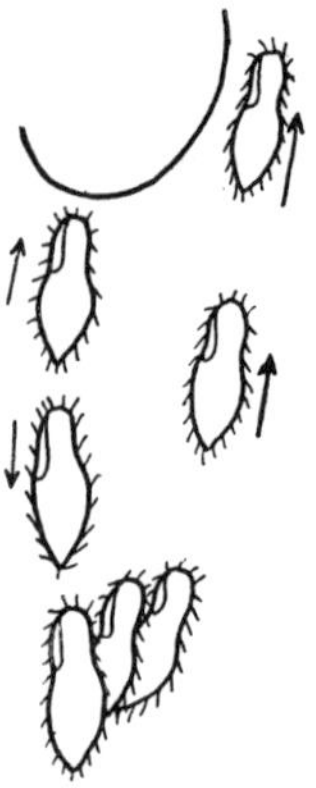

Avoiding reaction of Paramecium

What adventures befall a paramecium as it swims about in the waters of a pond? At all times it is acted upon by gravity, giving a negative response, that is, endeavoring to swim away from the bottom and toward the surface. If a solid object is encountered the paramecium backs off, swerves slightly and starts away in another direction, thus giving the avoiding reaction. When changes in the temperature of the water occur, the organism reacts most favorably toward regions heated to from 75°F to 82°F. A bit of decaying leaf may lie in its path, and the chemical products resulting from this vegetable disintegration, may attract or repel the animal. Thus paramecium is continually subjected to changes, reacting favorably as a rule to those that are beneficial and unfavorably to those that are harmful. It proves all things and holds fast to that which is good. No doubt internal stimuli, such as hunger, also influence the behavior of the organism.

Each species of protozoon reacts in its own peculiar way to stimuli. At first these reactions may seem fixed and the organism may be regarded as a helpless automaton, but protozoa are adapted to their environment both morphologically and physiologically or they would not be so successful in maintaining themselves. Their behavior must therefore be modified as a result of changes in the environment, and their automatic responses are not more automatic than the reflexes and instincts of higher animals. As Jennings says "if *Amoeba* were a large animal, so as to come within the everyday experience of human beings, its behavior would at once call forth the attribution to it of states of pleasure and pain, of hunger, desire, and the like, on precisely the same basis as we attribute these things to the dog."

In this chapter I have been able to refer to only a few of the many important contributions to the advancement of science resulting from the study of protozoa. They were rather late in putting in an appearance, but since they came "out of the everywhere into the here" in 1675, they have engaged the attention of thousands of the world's best biologists, and their influence on biological progress has been out of all proportion to their size. In the chapters that are to follow we shall deal mostly with parasitic species that live in lower animals and in Man. This special field has both scientific and practical aspects of considerable importance and hence has a bearing not only on the development of biology but also on the promotion of human welfare.

In concluding this chapter I wish to present an original poem —not the kind of poem I used to write during my college days, on the approach of spring, but a poem such as a gray-haired and feeble protozoologist might dedicate to the fascinating microörganisms that have been his constant companions during long years of toil and trouble. The title of this poem is Out of the Everywhere into the Here.

Where did you come from protos dear?
Out of the everywhere into the here.

When protos grow up, what do they do?
Each one divides, and then there are two.

Don't little protos long for sex?
You'd be surprised; they vex and vex.

What do amebas do in the soil?
Eat the bacteria and fertilize all.

Why does the water smell so evil?
Colonial flagellates have gone to the devil.

Do gay little protos live too in the sea?
Yea, brother, yea, a large quantity.

What makes the pyramids so strong?
Nummulite shells, throng upon throng.

How did the parasite habit evolve?
From free-living ancestors, that's easy to solve.

How did the protos become so prominent?
Mass reproduction rendered them dominant.

Why do the protozoologists weep?
The're too many protos and too little sleep.

Chapter II

PARADE OF THE LITTLE MILLIONS

WHEN we encounter an animal that is unknown to us we ask first what it is, then where it lives and, if our interest is sufficiently aroused, how it maintains itself and its race in competition with all the multitudes of other organisms and with the many visissitudes in its environment. These questions are comparatively easily answered now-a-days since scientists have been engaged for centuries in accumulating information about animals. The first students of the protozoa must have thought they were seeing things when there suddenly appeared under their recently-discovered microscopes an entirely unexpected menagerie of diminutive, but actively moving, wild animals that had to be labelled and pigeonholed. No doubt if the conservatives of that period had been consulted they would have opposed the study of protozoology as a blow to the existing order of chaos. The situation was seriously complicated by the inefficiency of the early microscopes. At that time it never occurred to anyone that these tiny organisms did not contain systems of organs, and the microscopes available were not powerful enough to demonstrate the absence of these organs. Even today if I were called upon to provide a layman with an example of chaos I could think of nothing more effective than a well-stocked culture of protozoa. A microscopic glance at such an assemblage would be like peeping through the keyhole at a milling throng of whirling dervishes.

Students of human evolution tell us that no obvious improvement in what we artfully call our minds has taken place within the past 2500 years, hence those zoologists who first undertook the study of the protozoa must have been as intelligent as we are today. If we were put on a spot equally hot we would probably also endow the common fresh-water flagellate *Euglena* with mammalian eyes and the ciliates with not one, but a whole collection, of little stomachs.

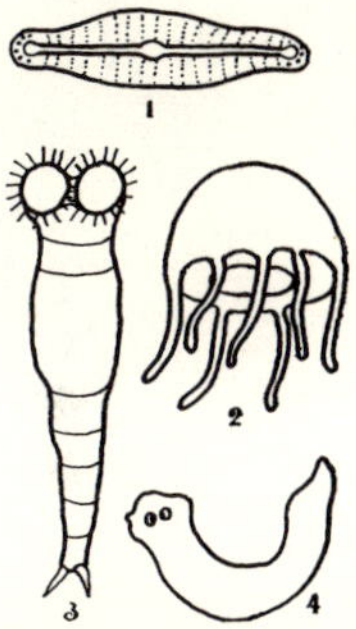

Former Protozoa. 1. Diatom. 2. Jellyfish. 3. Rotifer. 4. Flatworm.

CLASSIFICATION OF THE PROTOZOA

At first the protozoa were classed with other minute organisms including diatoms (one-celled plants), round worms (nematodes), wheel animalcules (rotifers) and the larvae of

FIG. 16. Do certain organisms belong to the plant kingdom or to the animal kingdom? The zoologist settles the question by flinging a bottle of ink at the botanical devil.

flatworms, coelenterates (jelly fish, etc.) and mollusks (oysters, clams, snails, etc.). But in course of time, as the true structure of the protozoa became known, they were gradually segregated as a distinct group in the animal kingdom, and are now studied by doctors, freshmen, and professors of protozoology. Even today, however, we don't know exactly where the protozoa begin or leave off, so gradually do they grade into the kingdom of plants on the one hand and into the many-celled animals (Metazoa) on the other. How some of them would chortle if they but knew the quantity of ink that has been shed by botanists and zoologists in their attempts to prove them either plants or animals! Certain of the more highly organized protozoa are undoubtedly multicellular and would be included among the Metazoa if we adhered strictly to our definitions of the various groups in the animal kingdom, but usually no difficulty is encountered in distinguishing these from the Porifera (sponges) and Coelenterata (jellyfishes, etc.) that comprise the two most primitive phyla of the Metazoa with which the protozoa might become confused.

As a result of the patient observations of the early biologists it is now a simple matter to separate the many thousands of different species of protozoa into four large groups on the basis of the presence or absence of locomotor organs and the nature of these when present. At first each species of protozoon discovered was isolated and described as a unique organism, but in course of time the method of nomenclature introduced by Linnaeus in 1758 was adopted for protozoa and the scientific name of the species is recognized throughout the world regardless of the language or the character of the government of the country to which the biologist using it belongs.

THE LIFE OF *Amoeba Proteus*

If we were to sit on a reviewing stand and watch the protozoa pass by, we would be greeted first with a battalion that progress by means of false feet (pseudopodia) and are known as Sarcodina, a term that means flesh-like. The leader and most common species belonging to this class is *Amoeba proteus.* How many young men and women have begun their study of zoology with a frantic search for this elusive member of the animal kingdom we don't know, but the number must reach into the hundreds of thousands. And how many artifacts have been mistaken for *Amoeba proteus* is also a matter for conjecture. The difficulty is easily stated; *Amoeba proteus* is only about 1/100 of an inch long and almost transparent, hence, unless large numbers are present, and the amount of light is cut down, and the observer is accustomed to the use of the compound microscope, usually a long exploration among the débris on the microscopic slide is required before the game of hide and seek is ended,—and not infrequently the instructor is finally called upon to demonstrate a specimen for the distracted student.

Amoeba proteus succeeds in maintaining itself and its race in this hectic and overpopulated world, and carries on the fundamental processes necessary for success that go on in our own

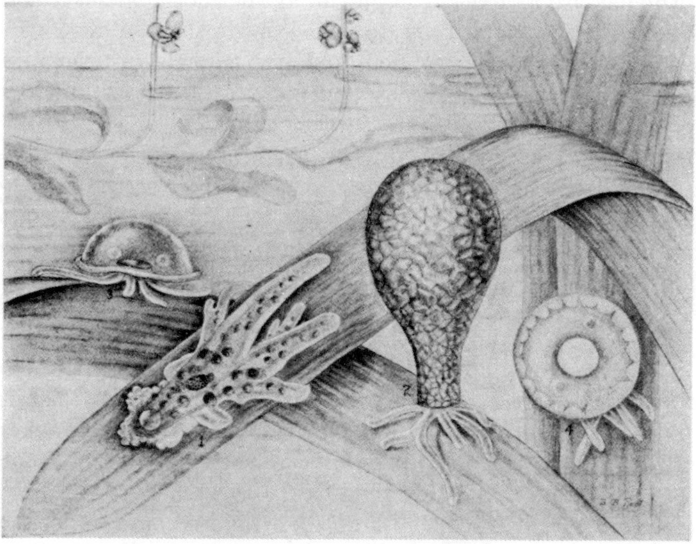

FIG. 17. *Amoeba proteus* (1) and its relatives *Diffugia* (2) and *Arcella* (3 and 4) are here seen taking a little stroll on a Sunday afternoon.

bodies, but its structure and mode of life are vastly different. There are no organs set aside for the performance of certain definite functions, such as circulatory, muscular, respiratory, excretory, nervous and reproductive systems and no well-developed sense organs such as eyes, ears and nose. All one sees is an irregular, colorless particle of "animated jelly" that is continually changing its shape by thrusting out and withdrawing little fingerlike processes, the pseudopodia. This animated jelly is protoplasm in which a close examination will reveal a disc-shaped, granular nucleus, a clear contractile vacuole, possibly other clear non-contractile vacuoles, and many food particles contained in food vacuoles. With this apparently very simple endowment, *Amoeba proteus* is ready to defy the rest of the living world, and does so with notable success. In other words, ameba is a jack of all trades and a master of every one of them.

Amebas are as fond of night-life as the most modern debutante. Furthermore, they do not sleep during the day, wake up just before the cocktail hour and arrive home with the milkman, but are active both day and night. Sleep was invented by someone (God bless him) further along in the scale of life. The ceaseless activity of ameba curiously enough derives no stimulus from sex since amebas are too young in the scale to know about the facts of life. What a marvelous method of simplifying existence, but how much they miss they will never know.

FOOD HABITS OF AMEBA

This constant and indefatigable industry is largely devoted to hunting, or perhaps, since amebas live in the water, fishing would be a better term. Unlike many human beings, however, amebas do not eat anything that the cook may place before them, but practice a rather rigid food selection. Other minute animals and microscopic plants are favored and especially certain common protozoa belonging to the flagellate and ciliate classes. You are at this point expected to wonder how it is possible for such a sluggish animal as ameba to capture its rapidly swimming relatives. The answer is that "slow but steady wins the race."

In moving from place to place, amebas thrust out pseudopodia and then flow into them, repeating the process as long as what they encounter proves to be good for them. They of course don't know where they are going, nor even where they have been, and, unfortunately for them, don't realize what a remarkably adventurous life they lead. However, they respond to stimuli of various sorts, such as changes in contact, light, temperature and the chemical constitution of the water. Thus when they approach a chilomonas flagellate or a paramecium ciliate that chances to be at rest, they slyly extend their pseudopodia without disturbing the victim who suddenly finds himself enclosed in an insidious trap from which there is no escape. The enclosed space becomes smaller and smaller and the captive acts as though paralyzed with fear, much like

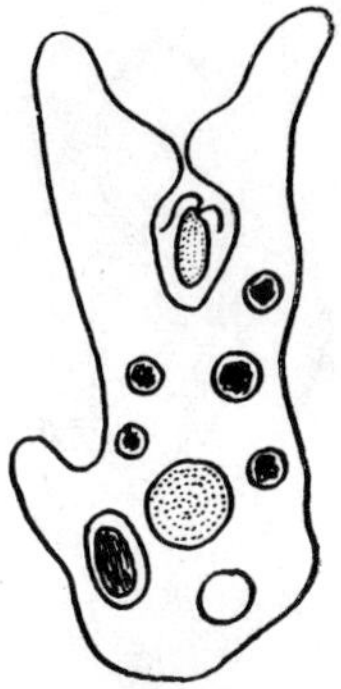

Ameba capturing a small flagellated protozoon

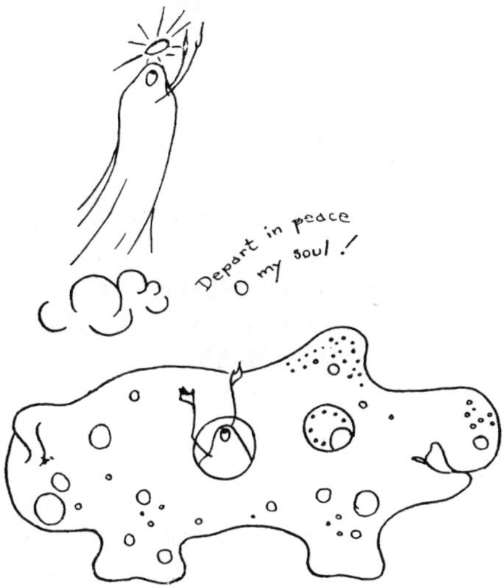

FIG. 18. Ameba engulfs its prey alive and soon the poor captive gives up its little ghost.

Poe's hero of the pit and the pendulum. Finally the food cup formed in this way is covered by pseudopodia pressing in from the sides, the food vacuole that results passes back into the body, the captive gives up his little ghost and the ameba prepares to digest a well-earned meal.

Small flagellates like chilomonas appear to be corraled by ameba with little difficulty, but paramecia that are almost as large as the amebic fisherman offer a problem that is solved in a fashion so improbable as well nigh to challenge belief. As before, the victim is approached so craftily that the trap is sprung too late for its escape. But no food vacuole is large enough to hold an entire paramecium, hence the ameba proceeds to cut the ciliate in two with its jelly-like pseudopodia, half a loaf being better than none. Just how the soft, elastic-looking pseudopodia succeed in bisecting the comparatively rigid paramecium is still a mystery unparalleled in the animal world.

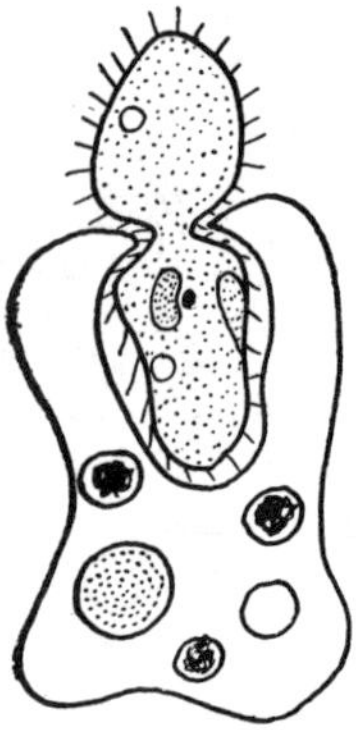

Ameba cutting a paramecium in two

Wonders continue after the food is "swallowed" since the apparently unorganized protoplasm secretes into the food vacuole digestive juices capable of dissolving the hapless prey, although it would not be surprising if a "stomach" ache were to develop in the vacuole containing the half-paramecium. One shudders to think of how he would feel after a meal comparatively as heavy as that. When the food is digested, it is absorbed into the protoplasm surrounding the vacuole, and since the vacuole is carried about by protoplasmic movements within the ameba's body, it is well distributed without the aid of a circulatory system. Assimilation follows, that is, the transformation of food into living substance. Indigestible particles pass out through the body wall usually at some point near the posterior end.

Considerable quantities of water are taken in when food vacuoles are formed, and water may also enter through the surface of the body. Such quantities of water would soon bring about death by dissolution were it not for the contractile vacuole which from time to time bursts at the surface and discharges liquids that accumulate in it to the outside. Amebas are thus rather unsanitary creatures; they "drink" the water in which they bathe and into which they discharge their waste products. If enough of them are confined in a small amount of water they may actually poison themselves with their own excretions.

The amazing energy necessary in order that amebas may continue what really amounts to perpetual motion, inasmuch as there has probably been no cessation of amebic activity since the evolution of the race millions of years ago, is furnished by the union of oxygen with the protoplasm. This chemical process results in compounds such as urea and carbon dioxide which are excreted by the contractile vacuole or through the general surface of the body.

BEHAVIOR OF AMEBA

Ameba moves away from a needle

Amebas seem so placid and sluggish that only by a long stretch of the imagination could they be called irritable; they certainly don't appear agitated when one views them under a microscope. Nevertheless, amebas behave like amebas because of this peculiar property. In the language of science

FIG. 19. Even amebas can't do as they please without suffering punishment but must behave like dutiful protozoa and obey the ameboid commandments.

irritability is exhibited whenever an organism responds to a stimulus no matter how weak the stimulus is nor how slow the response. The stimulus is usually a change in the environment. Thus, *Amoeba proteus* will respond by a negative reaction when touched with a needle, that is, it moves away as a result of a mechanical shock; it will migrate into a region where the water contains certain chemicals and away when other chemicals are present, such reactions probably accounting for food selection since different kinds of food give off different chemical products; it will move away from a strong light but toward a weak light; and it is sensitive to changes in the temperature of the water.

An ameba as it flows along over the vegetation in a fresh-water pool is being acted upon at all times by several of these stimuli. What the animal actually does depends not only on the relative strength of these various stimuli but also on its previous activities. A "hungry" ameba may respond to a certain change in the environment very differently from a well-fed specimen. Any number of experiences can be imagined which might modify the ameba's behavior, such as a strong light, a high temperature, contact with a moving object, and the presence of other types of protozoa.

FIG. 19 A. Ameba sheds a few tears when the doctor reveals its sexless condition.

Amebas apparently possess no sex appeal and stagger through existence without the most infinitesimal exhibition of love,—married, free or mother. It is difficult to forgive them for this omission, although it does make their life history more simple to understand. When *Amoeba proteus* reaches its full size it divides into two amebas. Each of these grows to the adult condition in about three days and then divides again. This appears to have been going on for millions of years and may continue until the end of time. However, if amebas were not so valuable to biology teachers as classroom material for students, all this energy and reproductive activity would hardly seem worth while.

RELATIVES OF AMEBA

Amoeba proteus is well provided with nephews and uncles and other more or less remote relatives. Family resemblances are so close, however, that only an expert amebiologist can distinguish one from another. We don't know why amebas leave home and evolve into distinct species, but the migratory habit has brought about the world-wide distribution of this type of animal life. Amebas seem to be everywhere, in bodies of fresh water, in ditches, in the soil in our door yards, in the fields, in swamps, in salt water, in most of the lower animals, and even

FIG. 20. The relatives of ameba, including its cousins and its aunts, all look plump and well-fed.

in human beings. Most of us are ambulatory banquet halls for the entertainment of amebas, but don't know it. This is perhaps just as well, since no one likes to think of the myriads of crawling things that may be roaming about among his insides. I am often asked how one can be a 100 per cent American when his bowels are swarming with amebas. Each of these species of amebas differs from every other species not only in visible structure but also in habitat and activities. We shall have occasion later to see what some of these amebas are like and how obnoxious they can be, hence we may now proceed to discuss some of their distant relatives.

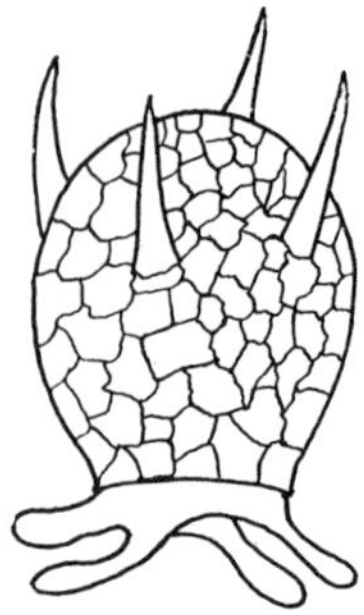

Difflugia builds itself a house of minute pebbles

Among the most conspicuous protozoa that a sample of débris scraped from a pond weed may reveal are ameba-like Sarcodina that are protected by a shell. Closely related to *Arcella* (see page 21) is a shelled type called *Difflugia,* a name derived from the characteristic flowing out of the pseudopodia through the opening in the shell. Like the third little pig, the ameba-like difflugia is willing to work hard to make its house impregnable, and, as a result, builds a shell of minute grains of sand which are glued together by a sticky secretion and form a firm house of stone, well suited to withstand attacks by any "big bad wolf" that may be seeking to devour it. Otherwise difflugia is similar to arcella, and, if it were deprived of its shell, would resemble a naked ameba. More than a score of species of difflugias have been described. Some are long and slender and others are short and stout; some have a round dome and others a pointed dome; and several bear spines on the dome or have teeth around the mouth of the shell.

Many other types of shelled protozoa that live in fresh water may claim relationship to arcella and difflugia and hundreds of species that construct elaborate coats of armor, such as globigerina and radiolaria (see Chapter I), live in the sea. The presence of a protective shell does not seem to make them any more successful in their struggle for existence than are the naked amebas, but at any rate their marvelously complicated and beautiful shells arouse our profound admiration.

A sun animalcule captures a small organism

Another type of the class Sarcodina that never fails to excite the astonishment of one who observes it for the first time is the sun animalcule. This is a sun that is as much smaller than the earth as our sun is larger than the earth. Its body seems almost too delicate to endure as it floats about in a pond or rolls along on the points of its rays over the water plants. From each little sun extend out hundreds of long, slender, pseudopodial rays. One would expect these frail-looking rays to collapse at a mere touch, but careful examination reveals a rigid axial filament in the center of each which guarantees firmness. These filaments are, however, capable of movement, since, when a minute animal swims against a pseudopodium, it may stick to the protoplasm surrounding the filament; then neighboring pseudopodia "come to the aid of the party" and bending over help force the prey down into the body where it is speedily killed and digested. If the sun animalcule were enlarged to the size of a lion its predatory habits would soon put it in the category with the ravenous beasts. Furthermore, its custom of "swallowing" its food alive is as cruel and unusual as our practice of eating oysters raw.

THE WHIP-BEARERS

One striking peculiarity of protozoa is their unpleasant appearance when a large collection of them is viewed from a distance and their great beauty when one or a few are examined under a microscope with high magnification. Thus the green scum that may form at the surface of a stagnant pool provokes a feeling of disgust, whereas the components of this scum, which are often flagellates of the genus *Euglena,* are clean-looking, cigar-shaped little organisms resembling streamlined submarines with curved, parallel grooves in the surface and oval bodies within that are colored green by chlorophyll. Euglenas are among the most common of the flagellated protozoa, or Mastigophora (which means whip-bearers), and perhaps the easiest to collect for study. They cannot be called typical animal flagellates because their food is largely manufactured within the body by a process known as photosynthesis, whereby starch is formed by the green chlorophyll from carbon dioxide and water in the presence of light. Furthermore certain envious botanists claim that euglenas are plants.

Euglena exhibiting euglenoid movements

The protoplasm at the surface of the flagellates is more rigid than that of the amebas hence changes in shape are less pronounced, although they do occur. Euglenas, for example, may swim through the water like a miniature fish, or may squirm about with contractions of the body resembling peristaltic waves; changes in shape of this type, wherever they occur among animals, are called euglenoid movements. But perhaps the most striking peculiarity of euglenas is the red eye-spot near the anterior end at the base of the flagellum. This eye-spot is not a complex organ like the eye of higher animals, but is a particle of protoplasm in which are embedded a number of granules of hematochrome (meaning blood-colored), a substance that has many of the characteristics of

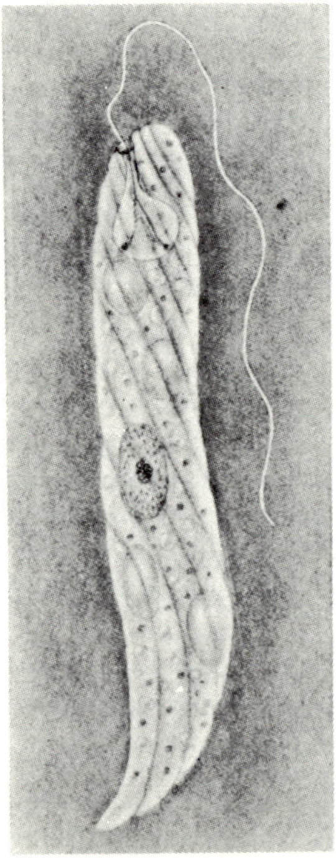

FIG. 21. Free-living flagellates such as *Euglena*, may look disgusting as green scum on a pond but are graceful and attractive when viewed under a microscope.

the pigments that occur in true eyes. Just in front of the eye-spot is a group of lens-like paramylum grains. At present we believe that these structures account for the fact that the anterior end of euglena is more sensitive to light than other regions of the body. This is fortunate for the euglenas since the anterior end is the first to enter a new environment and hence receives immediately any stimulus due to a change in the light.

If direct sunlight is encountered the euglenas will stop, then move backwards and start out in a new direction, thus exhibiting what is known as the "avoiding reaction." When given a choice of several light intensities, euglenas will select what to them is the "optimum"; this is the light best suited to them. Incidentally this very delicate sensitiveness to light must be of great survival value since the manufacture of starch and hence the nutrition of the euglenas depends on the presence of light. Whether or not euglena sleeps at night we don't know, but it might as well since it has no jacklight and cannot go night-hunting for other animals as ameba does.

VOLVOX AND THE APPEARANCE OF SEX IN THE ANIMAL KINGDOM

Among the near relatives of euglena are many species consisting of a number of individuals that are more or less intimately united to form colonies. Several of these colonial forms have already been described in connection with the tastes and smells that they impart to drinking water (see page 3). These species, as well as the euglenas, succeed in maintaining themselves in great abundance without the intervention of sexual processes. One of the colonial flagellates, however, has developed into a real many-celled organism and has acquired the remarkable phenomenon of sex. Furthermore sex in this flagellate is not as simple as one would expect in a creature so slow in the scale of life, but is comparatively complex. This organism is known as *Volvox*.

Volvox doesn't seem to have a care in the world but "just keeps rolling along" through the water of the pond in which it lives like a bouyant, animated baseball, hence the scientific name *Volvox* (from the Latin word volvere which means roll along). As our illustration shows, the volvox colony consists of an outer layer one cell thick enclosing a cavity filled with liquid; thus it really resembles a hollow rubber ball more than it does a baseball. Each of the thousands of individuals that make up the colony possesses two flagella, a nucleus, and an eye-spot, and is connected physiologically by protoplasmic strands with each of the six individuals surrounding it. Most of the members of the volvox colony are actively engaged in

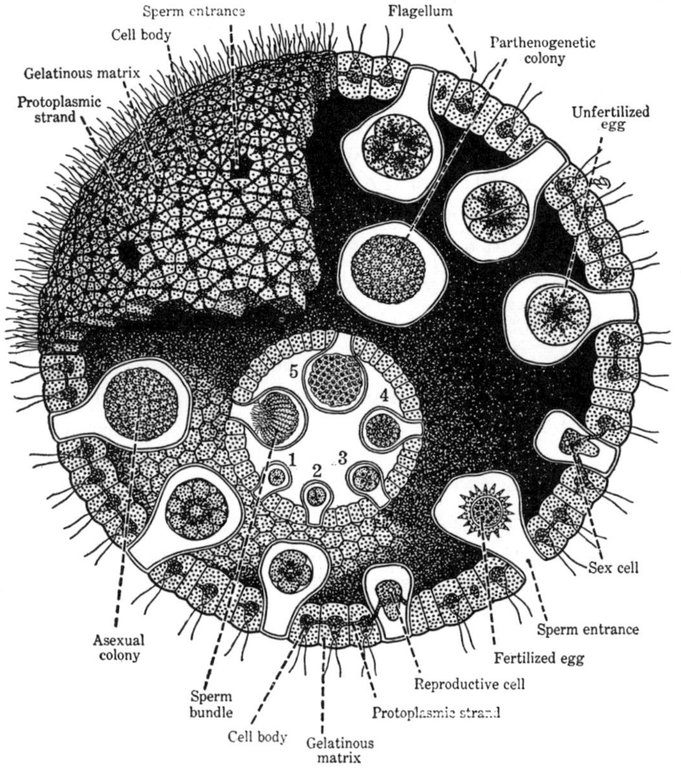

FIG. 22. The volvox colony consists of hundreds of similar "somatic" cells, and also both asexual and sexual reproductive cells.

propelling the organism through the water, in growing and in multiplying, thus increasing the number of members, but a few individuals develop into distinctive cells that are set aside for reproductive purposes. These germ cells, or sexual cells, are of three types whose histories are quite extraordinary.

First, a type of germ cell evolves that enlarges and undergoes division after division until it becomes recognizable as a distinct colony. This type of germ cell is called a parthenogonidium (from the Greek terms meaning virgin seed). New colonies that arise in this way eventually break out of the mother colony and live a free existence.

A second type of germ cell becomes much larger than its companions and because of its later history is referred to as a female cell, or egg. The third type of germ cell also grows large, and,

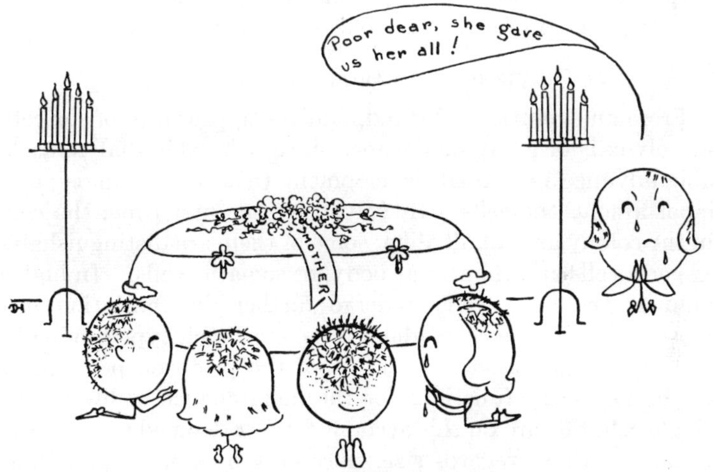

FIG. 23. Mother volvox sacrifices herself that sex may be introduced into the animal world; her sexually produced offspring weep about her bier.

as the result of many successive divisions, develops into a plate of very minute spindle-shaped cells, about 128 in number, each provided with an eye-spot and two flagella; these are male germ cells comparable to the spermatozoa of higher animals. Nature has furnished volvox with more male cells than female cells, which is no doubt fortunate for the continuation of the race since the male cells must seek out and fertilize the female cells before the development of the egg can proceed, and in this process most of the male cells are probably unsuccessful. Only one male cell unites with a single egg. Then this fertilized egg surrounds itself with a firm protective wall which it secretes, in which condition it passes the winter.

In the spring the eggs break out of their snug quarters and by numerous divisions build up an entirely new generation of volvoxes. But if you have tears, prepare to shed them now, because the parent colony that took so much trouble to develop a complex process of sexual reproduction falls to the bottom of the pond and death's bright angels remove its poor remains. This is the first advent of inevitable natural death in the animal kingdom, and all for the sake of sex. Is it worth it?

DIVISION OF LABOR AMONG CELLS

From an evolutionary standpoint the appearance of sex cells in volvox is of great significance since it heralds that remarkable advance in animal development that we call division of labor among the cells. In volvox for the first time, the cells in the colony are not all alike; some of them are distinguishable as germ cells and the rest as body, or somatic, cells. In higher animals the body cells undergo further differentiation into muscle cells, nerve cells, gland cells, etc., each type having its own particular functions. Most of the protozoa perform all of the necessary functions for the maintenance of life with a single cell, but are on this account forever doomed to a state of insignificance as regards size, since only by means of differentiated cells built up into tissues and organs can a large body carry on the many physiological processes necessary for continued existence.

OTHER WHIP-BEARERS

Euglena and volvox are flagellates that number hundreds of interesting plant-animals, or Phytomastigina, among their relatives (Phytomastigina, a term derived from Greek words meaning plants with whips). The rest of the flagellates are "animals with whips," or Zoomastigina. Like other animals the Zoomastigina depend on food built up by plants and this food is ingested by means of a cell-mouth or absorbed through the surface of the body. Among these are the intestinal flagellates described in Chapters V and VII and the blood-inhabiting flagellates discussed in Chapter VIII.

Among the more interesting animal-like flagellates that are easily obtained for examination is, in the first place, the "whip-bearing ameba," *Mastigamoeba*. Here is a clever ameba-like protozoon that captures and "swallows" food by means of pseudopodia, but finds ameboid movement too slow and hence travels with a long and active flagellum. Furthermore it has succeeded in adapting itself to two different habitats, the soil and fresh-water ponds, both of which offer almost infinite possibilities for expansion.

1. Mastigameba.
2. Mastigina

Many thousands of years ago, certain mastigamebas, apparently more adventurous than their playmates, took a trip with gun and camera through the alimentary canal of a frog tadpole and succeeded in setting up a residence in the intestine, hence now one may encounter their modified descendants in the rectum of frogs. To distinguish them from their ancestors, we call them mastiginas. They do not need to hustle as vigorously for their daily bread as do the mastigamebas, since they are actually suspended in food ready to be absorbed, hence their pseudopodia have become lazy and very seldom exhibit sufficient energy to accomplish anything worth while.

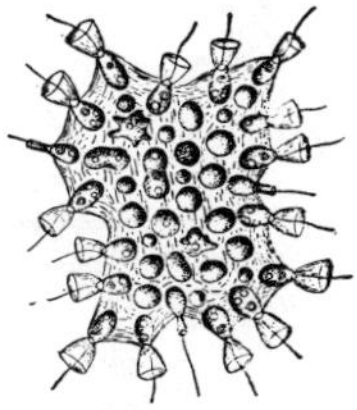

Proterospongia

Certain animal-like flagellates have, like volvox, adopted the colonial manner of life. One of the most intriguing of these is a species belonging to the choanoflagellates which has been named Proterospongia because it resembles what the ancestors of sponges might have been like (*Choana* means funnel and *proteros* means before). In the proterospongia colony two distinct types of cells are embedded in a jelly-like matrix. The inner cells are ameboid in appearance, but the cells at the surface are all fussed up with a stiff collar from the center of which projects a twinkling flagellum. These collar cells are also characteristic of sponges and no particular strain on the imagination is necessary to picture the evolution of proterospongia into a simple sponge, although many centuries elapsed before this simple sponge evolved into the maker of the skeleton that is kept in every respectable family's bathroom; or if not there, at least in the garage.

Paramecium swims **through** the trackless **water** in a spiral course **but the** axis of this spiral is a straight line.

CILIATED PROTOZOA

The ciliates among the protozoa are comparable to the birds among the vertebrates, that is, they always attract attention and are easier to observe than other types. Everyone who has ever studied biology knows that the slipper animalcule, paramecium, is "teacher's pet." The reason is not far to seek, since almost every infusion of pond weed in water sooner or later is found to be well populated with this large representative of the class Infusoria. We call the members of this class ciliates since they are more or less completely covered with minute hair-like cilia (from the latin word *cilium* meaning eyelid, and hence eyelash). These cilia are more useful to paramecia than our hair is to us, since, by their movements, they propel the little animalcules through the water, keep the surface of the body free from débris and help to drive food particles into the cell mouth. Furthermore cilia don't fall out almost as soon as they grow in, the way our hair does, so no one has ever seen a bald-headed paramecium.

If we were to compare the accomplishments of lower animals with those of man we would often be surprised to find that even such an insignificant organism as paramecium can do some things that we cannot. Perhaps paramecium is most conspicuously superior to us in its ability to move in a straight line although errors in any direction are possible. Human beings, on the other hand, walk around in a circle, if no land marks are available as guides, in spite of the fact that mistakes are possible in only two directions, right or left. Paramecium's achievement is made possible because it revolves on its longitudinal axis and continually swerves to one side as a result of the stronger beating of the cilia in the oral groove. The result is a spiral course, but, unless stimulated by some change in the environment, this spiral course will have a straight line as an axis.

Paramecium is obviously more highly organized than the

amebas and flagellates. In its life cycle is included a complicated process of sexual reproduction. How a paramecium attracts a mate is still an unsolved problem; the sexual urge appears periodically resulting in conjugation. However,

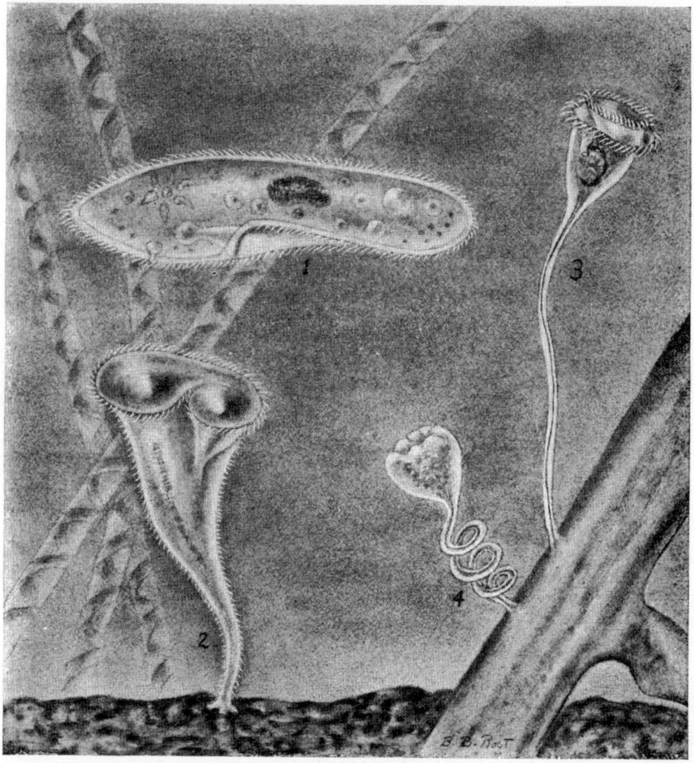

FIG. 24. Three of the most common species of ciliated protozoa are *Paramecium* (1), *Stentor* (2) and *Vorticella* (3). *Vorticella* may contract on its stalk and draw in its bands of cilia (4).

paramecium is not as carefree as one might at first suspect because it is well prepared to protect itself from certain other organisms in its environment. Just beneath the surface of the body is a layer of spindle-shaped cavities filled with a

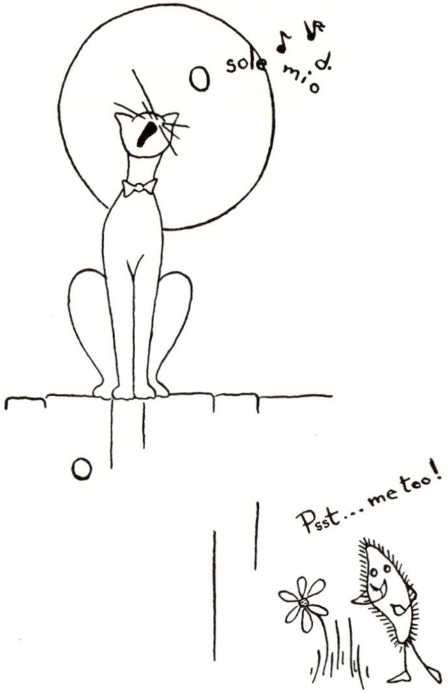

FIG. 25. Paramecium also feels the sexual urge but (fortunately) is unable to announce it vociferously to the world.

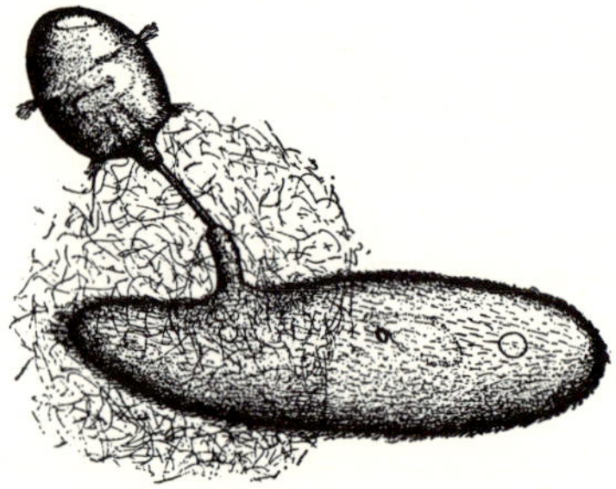

FIG. 26. Paramecium, when attacked by its enemy, didinium, explodes its trichocysts and drives its opponent away.

semi-liquid homogeneous substance. When properly stimulated this material is expelled through minute pores, and hardens into long threads in the water; these threads ward off the attack of the enemy.

GROUPS OF INFUSORIA

The Infusoria are separated into groups according to the character and arrangement of their cilia. Paramecium and opalina (see page 74) belong to a group that are provided with a complete coat of cilia, mostly of equal length. In a second group the cilia may be absent from certain regions of the body, but a zone of large cilia is present around the cell mouth. The large protozoan parasite of man, *Balantidium coli,* is of this type. The free-living ciliate known as stentor is a common fresh-water example. This trumpet-shaped species usually attaches itself at one end to some solid object in the water and is content to remain quietly at rest while the large cilia that are spirally arranged around the outer end create a current in the water that draws a supply of food particles into the mouth. Occasionally stentor responds to a craving for greater freedom by detaching itself and swimming about for a while in the boundless deep.

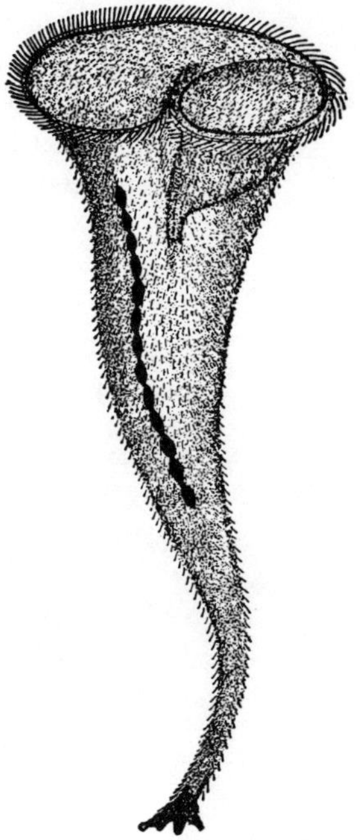

Stentor, the trumpet animalcule

A third group of Infusoria are likewise able to swim about when they so desire, but have adopted the creeping habit and ordinarily "walk" about on the leaves of water plants with "legs" consisting of clumps of cilia that have fused together to form large, powerful cirri (from the Latin word *cirrus,* meaning a tuft of hair). The body of these protozoa has become flattened in conformity with this mode of life. Some species have formed the habit of creeping about on certain fresh-water animals and have thus become parasites.

Another group of Infusoria are of the vorticella type. These animalcules are shaped like a bell, but are fastened upside down at the end of a stalk five or six times as long as the body,

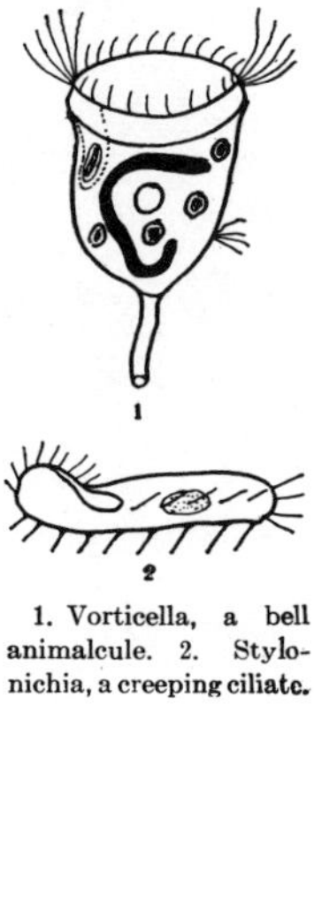

1. Vorticella, a bell animalcule. 2. Stylonichia, a creeping ciliate.

and the bells do not ring. The stalk is attached to some solid object, and from time to time contracts in the form of a coil spring thus drawing the bell backward. It then straightens out again, the bell being pushed forward in another direction. Around the outer end of the bell is a row of cilia that produces a vortex in the water directed toward the mouth opening. This brings food particles where they can be sucked into the

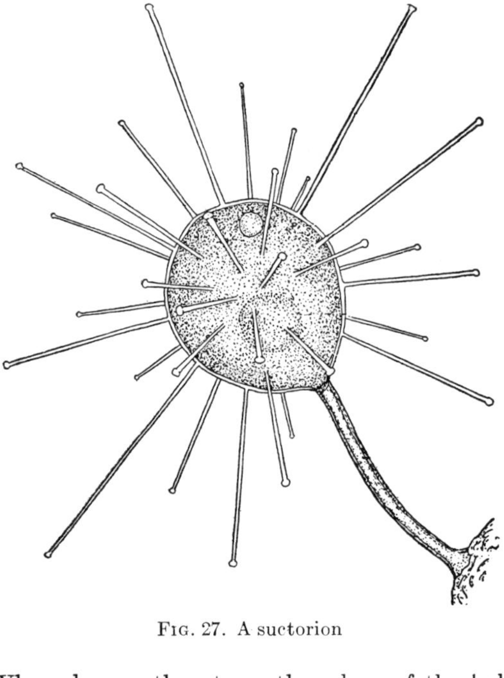

FIG. 27. A suctorion

gullet. When danger threatens, the edges of the bell may be drawn in over the mouth.

One species of bell animalcule was the first protozoan ever seen by man; it was discovered by van Leeuwenhoek in April, 1675. Some of the bell animalcules are colonial in habit, living in large groups, each on the end of a stalk; but the stalks are attached to one another somewhat like the branches of a tree. It is a familiar habit of these little creatures to attach themselves to the body of some larger animal in their aquatic habitat, such as a tadpole, snail or turtle. They have no thumb with which to indicate in which direction they wish to be transported, but they are benefited no matter where they are carried since new feeding grounds are certain to be encountered during the travels of their animated submarine. Several close relatives of the bell animalcule have adopted a

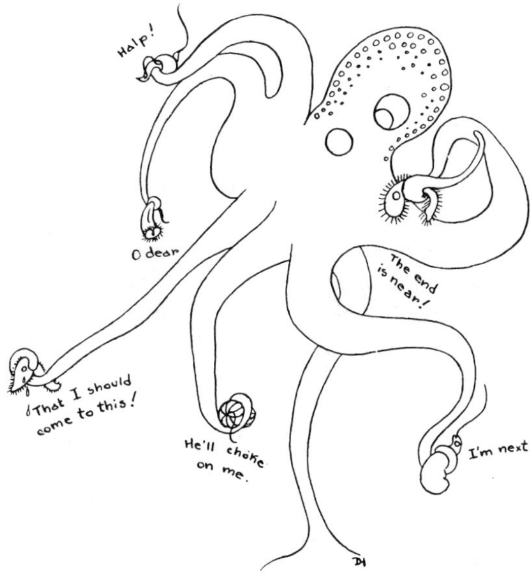

FIG. 28. Certain ciliates develop into suctoria with tenacles, which, like those of a ravenous devilfish, capture other helpless little protozoa.

parasitic habit and may be found creeping over the bodies of tadpoles and other fresh-water animals (see page 65).

This brings us to the last group of the Infusoria,—those peculiar predatory creatures, the suctoria, that remind one of miniature devil fishes. The birthrate of suckers among human beings may seem high, but among protozoa millions of suckers are continuously breeding more suckers in an endless array. These suckers are really ciliates that have cilia when they are babies but lose them as they grow up. In their place appear sucking tentacles with which they capture other minute protozoa, and by means of which they suck the living substance of their prey into their own bodies. Not a very pleasant occupation, but very successful, because sucking protozoa are very numerous not only in fresh water and in the sea, but as parasites in and on other animals. A continuous struggle is being carried on between each little animalcule and the other little animalcules in his neighborhood and only the fortunate few survive. The survivors are, of course, the best fitted for their mode of life.

SPORE-BEARERS

The last battalion of protozoa all belong to that fascinating, but disreputable, group of animals that we call parasites. Many of these produce at some stage in their life cycle minute resistant bodies known as spores. On this account the entire assemblage has been given the name Sporozoa which means spore-bearing animals. Some of the sporozoa are among man's worst enemies, for example, the malaria parasites (see Chapters IX and X). Others are pathogenic to lower animals and thus of great economic importance to man. Certain types of sporozoa are referred to elsewhere in this book, for example, the coccidians that are so destructive to rabbits and chickens (page 78), the piroplasmas of Texas fever in cattle (page 72), the myxosporidians of fish (page 68), and the microsporidians that destroy silkworms and honeybees (page 70).

Almost every species of higher animal is parasitized by its own peculiar types of sporozoans. These always injure their host since they penetrate the intestinal wall, muscles, blood cells or other parts of the body. Perhaps the easiest species for the amateur microscopist to secure are those that live in the reproductive organs of the earthworm and in the intestine of grasshoppers and mealworms.

To observe these you must first capture your earthworm or insect and "bring 'em back alive." The earthworm's body consists of a series of rings. Count these rings beginning with the anterior end until you reach the twelfth ring. Make a slit in the body wall at this point and several whitish bodies will protrude; these are reproductive organs. If some of the contents of these are placed on a microscopic slide and diluted with water the chances of finding the spores, and perhaps other stages in the life cycle of a sporozoon, are quite good. The accompanying illustration will give you some idea of what to look for. The most common stage is apt to be a cyst full of spindle-shaped spores.

A cyst of a sporozoon (Monocystis) containing spores.

Perhaps most of my readers would rather operate on grasshoppers than on earthworms. If so, a very interesting type of spore-bearer is available for inspection; it belongs to a gregarious order, the Gregarinida. The accompanying illustration indicates what to look for in the intestine. Here two gregarines are seen, each consisting of a small anterior segment and a larger posterior segment. One gregarine is attached by its anterior end to the posterior end of the other,—a condition known as syzygy (pronounced sī-zī-gee). Other individuals of various sizes may also be found that are not so gregarious but, like Kipling's cat, wander about "by their wild lone." Their method of locomotion that makes their nomadic life possible is still a biological enigma, since they glide slowly about without any obvious means. In most respects the life cycle of this gregarine is much like that of the monocystis in the earthworm. Spores are formed which pass out of the grasshopper's body in its excreta. If these are swallowed by other grasshoppers, they hatch in the intestine, and, after living for a time within the cells of the intestinal wall, break out into the lumen of the intestine and glide about seeking a mate, and finally a new generation of spores are formed. Life for gregarines is just one generation of spores after another.

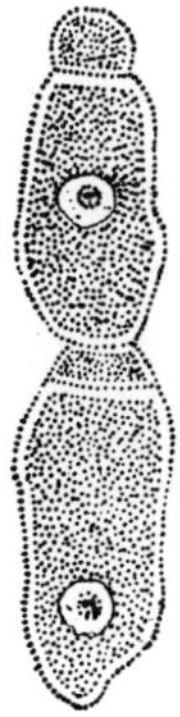

Two gregarines from a cockroach, attached together in syzygy.

We do not contend that this chapter provides a very adequate review of the entire protozoan world,—only a five foot shelf of books could properly present what is known about the thousands of species that have been described,—but we hope our illustrations and descriptions will help to elucidate the activities of the parasitic types that are the principal subjects dealt with in succeeding chapters.

The enormous numbers and varied characteristics of these minute animals inspired the following poem.

Forward little millions,
Marching gaily on,
Little naked 'meba,
Leads the mighty throng.

Comes the green Euglena,
And Paramecium too,
Followed by the Spore Bearers,
Cavorting two by two.

Onward little millions,
Marching for to see
If they cannot dominate
The world, and you and me.

Chapter III

AND LITTLE FLEAS HAVE LESSER FLEAS

THE term Parasite when applied to human beings has much the same meaning as that recognized by biologists. Originally it was used for persons who frequented the tables of the rich and earned their welcome by flattery. Parasitism in a restricted sense is now used only for the type of association between two different species of organisms in which one species, the host, is actually injured by the other species, the parasite. A parasite of this type is said to be pathogenic to its host. In a broad sense Parasitism implies any condition in which one species lives in or on another species. One remarkable type of association is known as symbiosis, a term derived from a Greek word meaning "living together." This term is now ordinarily employed only for partnerships that have become of the Damon and Pythias pattern in which the partners are mutually beneficial and often are unable to live apart.

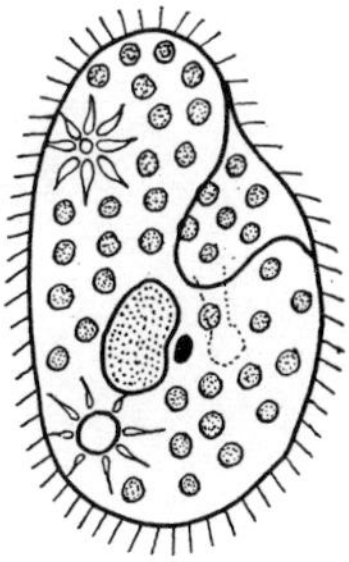

Paramecium bursaria, a ciliate whose body is filled with green plants (algae).

SYMBIOSIS AMONG PROTOZOA

Symbiotic relationships exist between plants and animals as well as between different species of plants and different species of animals. Thus many protozoa may contain unicellular plants, with which they live in perfect harmony. One of the species of paramecium, *Paramecium bursaria,* is usually colored green by algae of the zoochlorella type. These algae are ingested by the paramecium, just as are solid particles of food, but for some unknown reason are not digested. On the contrary they grow and multiply within the protoplasm of the paramecium. When the protozoon divides each daughter paramecium carries with it a certain number of the algae and in this way the association may be continued indefinitely. The alga may be designated the photosynthetic partner since it manufactures food from inorganic materials in the presence of light, a process known as photosynthesis. The protozoon protects the alga with its body, gives it a continuous free ride through the pathless waters, leads it into lighted areas favorable for the process of photosynthesis and provides it with carbon dioxide and nitrogen. The alga pays for these benefits

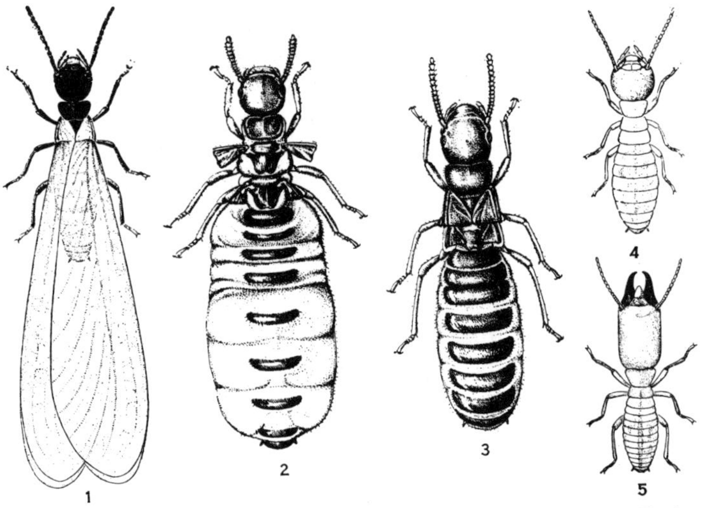

FIG. 29. Termites. 1. Type with wings. 2. Queen. 3. Male. 4. Worker. 5. Soldier.

with a constant supply of oxygen, with an occasional meal of carbohydrates and with efficient sanitary service which relieves the protozoon of nitrogenous wastes. Protozoa that live in fresh water are most often invaded by zoochlorellas, whereas marine species usually live in symbiosis with brown-colored algae known as zooxanthellas.

One of the best examples of symbiosis known is that of termites, or white ants, and their intestinal flagellates. For example, the intestine of one large species of termite, *Termopsis angusticollis,* is swarming with four different types of flagellates. This termite feeds on wood. If it is kept in pure oxygen for 72 hours at a pressure of one atmosphere, all of these protozoa are destroyed, but the termite is unharmed. When returned to a normal atmosphere, the termite will eat wood, but is unable to digest it without the aid of the protozoa, and dies of starvation within from three to four weeks in the midst of plenty. If a termite defaunated by oxygen is reinfected with protozoa from another termite its digestive powers are restored to it and it lives on indefinitely. If a termite is starved for six days it loses one type of flagellate, *Tricho-*

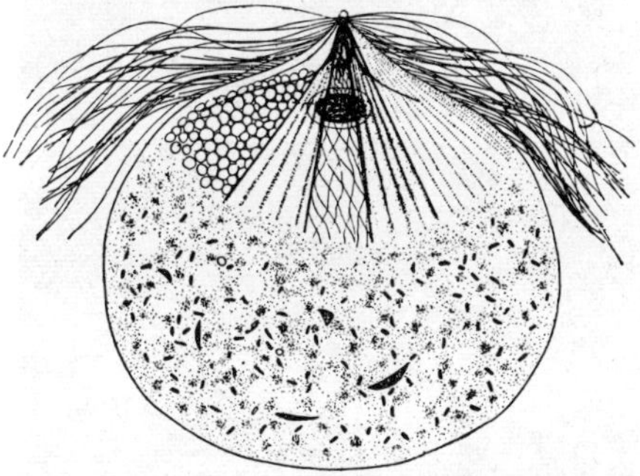

FIG. 30. This is a sketch of the most important intestinal flagellate (Leidyopsis) of the termites. The small rod-shaped bodies are particles of wood.

nympha; if it is starved for two days longer, it loses a second type of flagellate, *Leidyopsis;* if it is subjected to oxygen at a pressure of one atmosphere for twenty-four hours, it loses a third type, *Trichomonas;* and if it is subjected to oxygen at a pressure of one and one-half atmospheres for seven hours it loses the fourth type, *Streblomastix.* By various methods of treatment we have found that the termite will die in from three to four weeks if all protozoa are removed or if all but *Streblomastix* are removed; that it will die in about ten weeks if both *Trichonympha* and *Leidyopsis* are removed, but will live on indefinitely with any combination of flagellates if *Leidyopsis* is present or even with *Leidyopsis* alone. Obviously, this species of termite can exist only when the symbiote *Leidyopsis* is available to digest the wood it eats. *Leidyopsis* is equally dependent on the termite since it can live no where except in the intestine of this insect. The other three types of flagellates appear to be thumbers, bumming a joy ride at the expense of their hard-working relative *Leidyopsis*. If I were a termite I would hasten to emulate Romeo and do my best,

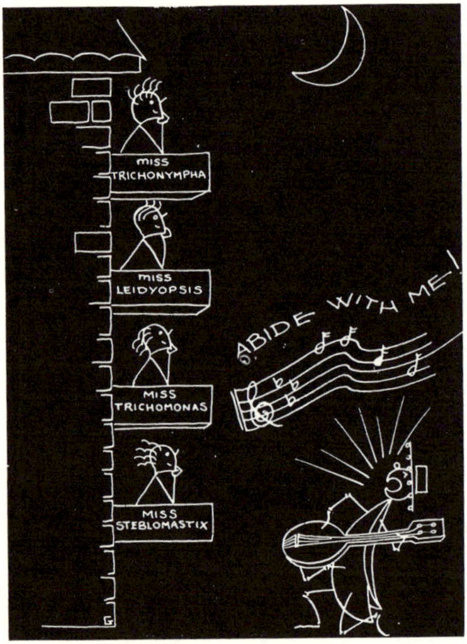

FIG. 31. This termite is taking the part of Romeo and trying with the help of sweet music to charm a few flagellated Juliets down from their balconies.

with the aid of sweet music, to charm a few flagellated Juliets down from their balconies.

Incidentally the most perfect method of transmission of protozoa from one host to another is exhibited by termite flagellates. The young termites soon after they hatch from the egg are fed by their parents on what is known as proctodeal food. This is excreta of the adults that comes from the posterior end of the alimentary canal, a part named the proctodeum. It consists of partially-digested cellulose and many flagellates. One hundred per cent of the young termites are therefore infected as soon as they are born, and are thus supplied with a complete collection of flagellates to digest their food for them.

COMMENSALISM

Certain protozoa, like the human parasites of olden times, frequent the banquet halls of the rich but are called commensals, a term that means living at the same table. So far as can be determined these species are of no benefit to their hosts; neither do they cause any obvious damage. They live on or within their host for the sake of the scraps of food they can steal from the pantry. Several species that live in the human intestine are of this type and have very appropriately been designated "food robbers."

Symbionts and commensals are of great scientific interest and a host is certainly lucky whose guests are content to live and let live, or willing to make only an occasional raid on the ice box. The real bad boys are the true parasites and when they are bad they are very horrid indeed.

PARASITES THAT LIVE IN PROTOZOA

No animal appears to be too small or too low down in the scale of life to escape being parasitized. There is some question regarding what species of animal belongs at the foot of the evolutionary tree, but amebas are certainly serious contenders for this doubtful honor. Various species of amebas including those parasitic in man, are invaded by an organism known as *Sphaerita*. *Sphaerita* is classified among the lower plants.

It is simple enough to distinguish a tree from a horse but many microscopic organisms resemble both plants and animals in structure and activities. Some biologists try to sidestep the dilemma by lumping these primitive organisms into a single category and calling them all Protista, a term meaning "the very first"; but most of us try to segregate them on the

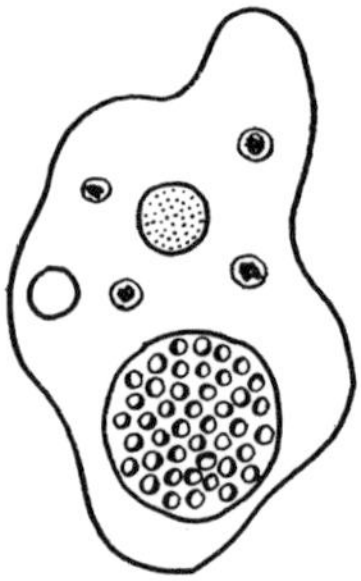

This ameba is parasitized by a sphaerita full of refringent granules.

basis of certain more or less constant characteristics. Thus the animal-like species, or Protozoa, usually possess well-developed locomotor organs, respond to stimuli quickly, lack chlorophyll and excrete waste products consisting principally of carbon dioxide and urea. The plant-like species, or Protophyta, usually possess less poorly developed locomotor organs or none at all, respond slowly to stimulation, are provided with the green coloring matter, chlorphyll, and excrete oxygen as well as carbon dioxide.

Thus plants manufacture food from non-living materials and animals do not; this means that all animals, even protozoa, can live only because of the labor of plants,—a sort of capitalistic system that has prevailed in nature ever since life has existed on the earth. If we accept these criteria, then *Sphaerita* is undoubtedly a plant. Infection of protozoa with sphaeritas is due to spores which penetrate the body, grow at its expense and then break up into many daughter spores. When alive, sphaerita resembles a sphere filled with refringent beads; when stained, these bead-like spores become deeply colored. Parasitic amebas and other parasitic protozoa, as well as free-living species, may be parasitized by sphaerita and other parasites, a condition known as hyperparasitism.

Amebas do not take their humiliating condition as hosts for microscopic plants in a prone position, but apparently look about in search of some evil for their own idle hands to do. Some of the amebas have learned to invade other protozoa, for example, certain opalinid ciliates that live in the rectum of frogs and toads. Once an ameba has succeeded in boring its way into an opalinid, it has reached a haven such as most people dream about but never live to realize. Free transportation, a constant food supply, and plenty of space in which to gad about and reproduce are provided. Sometimes one opalinid may contain over one hundred amebas. Then when the opalinid multiplies by binary division, each daughter opalinid has to accept her share of the nonpaying ameboid guests, and thus the offspring of the amebas inherit a comfortable and fully stocked residence generation after generation.

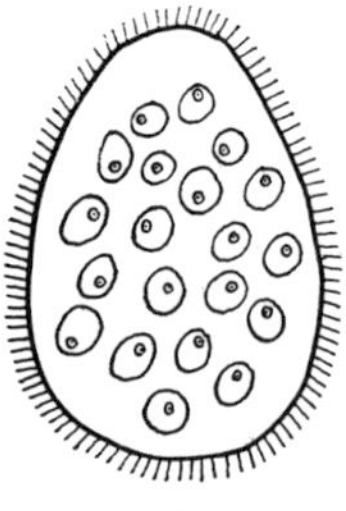

An opalina ciliate parasitized by amebas.

And Little Fleas Have Lesser Fleas

Amebas and opalinids belong to different classes of protozoa, but sometimes members of the same class prey upon each other. Thus certain ciliates of the suctoria type burrow their way into other ciliates of the paramecium type. Here, protected from all outside injurious agents, they thrive and multiply at the expense of their obliging host. The young

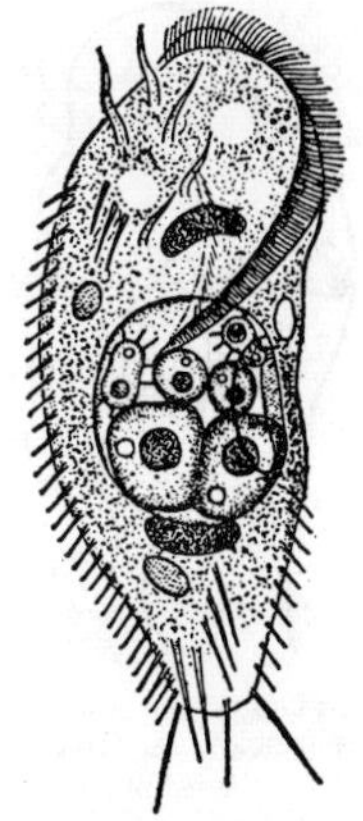

A ciliate parasitized by suctoria. The brood pouch is full of young "suckers."

FIG. 32. The birth of quintuplets would represent abysmal failure in the suctorian family.

suctorians are continuously rocked in the cradle of the deep as the paramecium swims about. Fortunately for them, their parents leave a passage through which they can escape to the outside. The free-living period in the existence of these young suctorians is spent in house-hunting for another paramecium in which new families can be reared. No birth control is practiced among the suctoria, and a new sucker is born practically every minute.

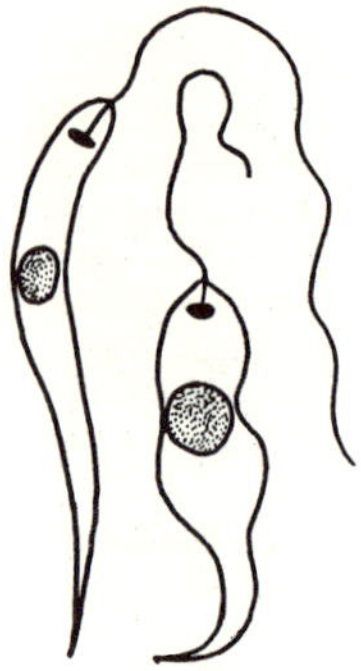

Flagellated protozoa that live in the latex of the milkweed plant.

PROTOZOAN PARASITES OF PLANTS

When a plant parasitizes a protozoon we are apt to be mildly interested but when a protozoon parasitizes a plant, that is news worthy of a scare-head. As already stated, animals cannot exist without plant food, but as a rule parts or all of the plant are eaten outright. In exceptional cases protozoa may actually live within the plant without seriously injuring it. This relieves the animal of the necessity of continually

FIG. 33. Milkweed bugs carry on an efficient taxi service for the flagellates that live in plants.

searching for new host plants on which to feed, since one plant serves as a satisfactory residence for the parasite as long as it lives. Of course, only very small animals can smuggle their way into the leaves and stems of plants, hence, as one might expect, the animal parasites of plants belong among the protozoa.

Flagellates appear to be the most frequent plant parasites. Latex plants are the type of vegetation selected in which to live. Latex is the milky contents contained in the cells of certain plants, such as our common milkweeds. These plants are abundant in the United States and many of them are infected, but our American biologists allowed the Frenchman Lafont to make the first discovery of these parasitic flagellates in a species of spurge on the island of Mauritius. When it occurred to us in 1924 that we might find similar flagellates in the vicinity of Baltimore we secured an abundant supply without difficulty from milkweeds. They were also found in plants at Tela, Honduras. Later I looked for them in the Philippines and procured a supply from plants growing on the campus of the University in Manila.

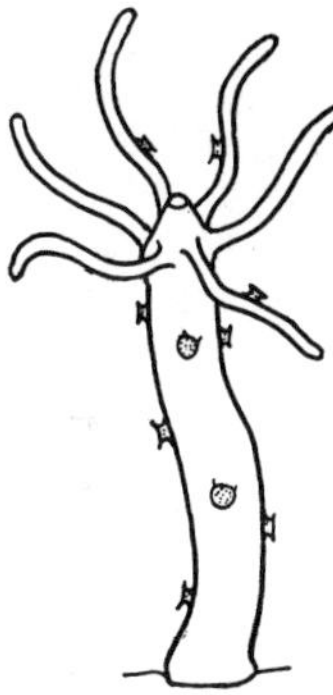

A fresh-water hydra with trichodina ciliates creeping over its body.

Obviously these flagellates cannot migrate from plant to plant without assistance. Their means of transmission was easily established, however, since latex plants are regularly visited by sucking bugs. When these bugs were examined it was found that as the latex is drawn into the digestive tract of the bugs, the flagellates become located in the salivary glands, where they undergo multiplication. Then when the bugs visit other plants, a bit of saliva containing flagellates may be injected into the latex cells thus infecting the clean plants. The bugs thus carry on an efficient taxi service from one plant to another. Whether infected plants and transmitting bugs are injured by the flagellates has not been definitely determined. At any rate it seems quite certain that the flagellates are not doing the plants any good.

ECTOPARASITIC PROTOZOA ON HYDRA AND TADPOLES

Those of you who have studied zoology are acquainted with the fresh-water coelenterate named *Hydra*. Frequently hydras are lousy with a species of ciliated protozoa known as *Trichodina*—a term that means hairy louse. The trichodinas creep about over the surface of the hydra. One of the peculiar characteristics of coelenterates, such as hydras, sea-anemones, and coral polyps, is the presence of stinging organs, called nematocysts. These are borne on wartlike projections which contain large and small stinging capsules resembling a battery consisting of large cannons surrounded by small machine guns. When a little animal swims within reach of a hydra these stinging capsules may explode and a hollow tube is shot into the victim through which is forced a paralyzing poison. In some unexplained way the trichodina ciliates as they creep

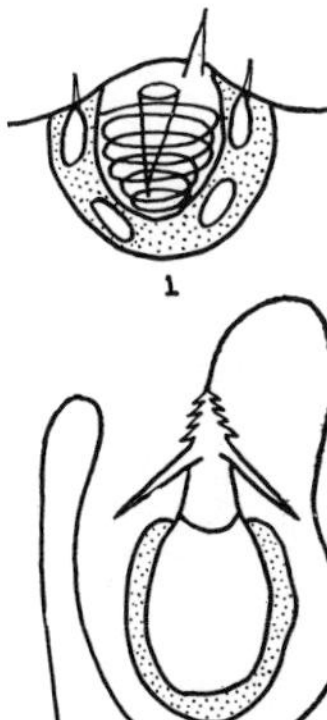

1

2

Stinging cells of hydra. 1. Unexploded. 2. Exploded.

about on the hydras fail to stimulate the explosion of the stinging capsules and thus escape injury from these otherwise very effective weapons. On the other hand, the hydras are seriously injured by their louse-like companions. Whether they are overcome by an itching sensation that they are unable to scratch, or are merely tickled to death remains to be determined.

Tadpoles are likewise embarrassed by trichodinas, but tadpoles develop into frogs, and the frogs leave the water, thereby

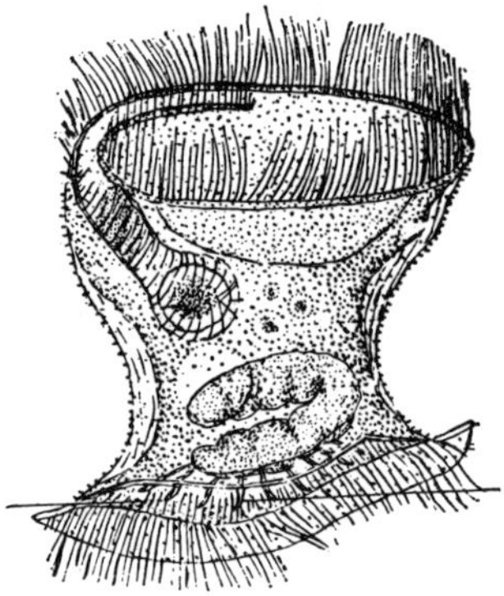

FIG. 34. Trichodina, an ectoparasite that creeps about on the surface of aquatic animals.

exposing the trichodinas to the air. This behavior must drive the trichodinas frantic. Some of them probably enter the mouth of the baby frog and are promptly swallowed and digested, whereas others, that may or may not be considered more fortunate, according to ones taste regarding such matters, probably enter the back door and life is spared to them. But who wants to live in the urinary tract of a frog. The trichodinas that do, have become somewhat modified for this type of existence and undoubtedly illustrate the evolution of an external parasite into an internal parasite,—but at what a price!

PARASITIC CILIATES IN EARTHWORMS

I have always been a strong proponent of back-yard biology, that is in the study of plants and animals that occur close to one's home. A few years ago one of my students wanted a little problem to work on so I suggested that she find out whether earthworms in this country are parasitized by intestinal ciliates. That evening I dug up a few earthworms in my back yard and the next morning she found that they were all infected. We then proceeded to organize a comprehensive program for the study of earthworm ciliates from various

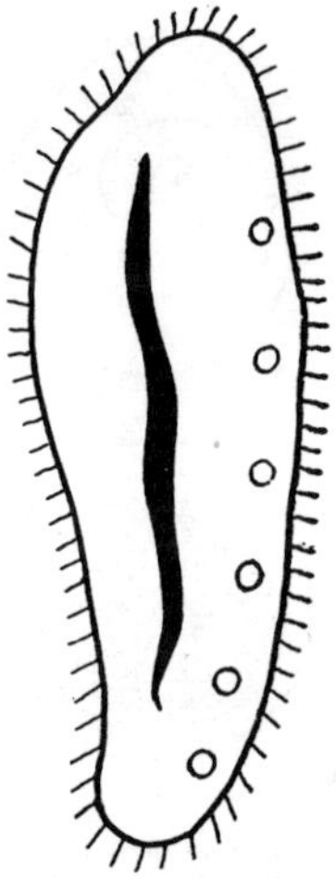

An earthworm ciliate

FIG. 35. Imagine the surprise of a protozoologist who finds parasites only in the earthworms that live in his back yard.

parts of Baltimore and from different regions in North America. We were soon deluged with a flood of earthworms but apparently the only earthworms in the New World that contained ciliates lived in my back yard. I don't suppose there is a better place for parasitized earthworms to live than in the back yard of a protozoologist but such a phenomenon certainly deserves public recognition.

PROTOZOAN PARASITES OF FISH

The only people who worry about the troubles of these lowly animals that are parasitized by protozoa are the biologists.

The problems involved are similar to those that still puzzle investigators who delve into the mysteries of human parasitic protozoa. But the chances of learning something of general biological significance are greater when the comparatively simple association between the lower animals are studied than when the enormously more complex relations between man and his internal menagerie are investigated. Almost everyone, however, is interested in fish, and when a protozoon attacks the pet fish in our aquarium or the fish that might otherwise serve to satisfy our hunger, the first reaction is to call on a biologist for expert advice.

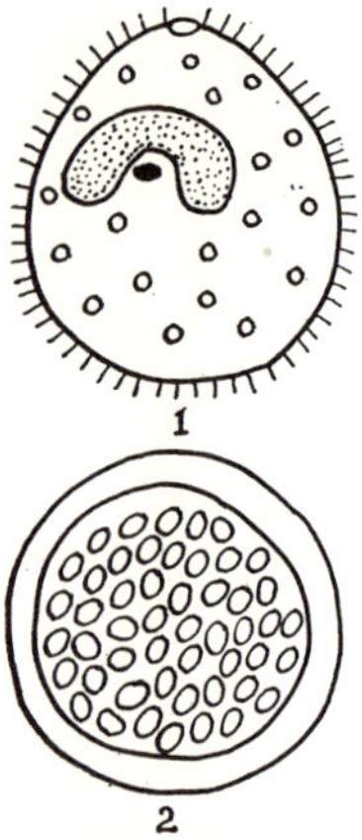

A fish parasite. 1. Active stage. 2. Cyst.

One of the most conspicuous external parasites of fish succeeds in existing although burdened with the scientific name *Ichthyophthirius multifiliis*. This ridiculously long term brings to mind the remark of a little girl to whom I was showing the wonders of animal life in the Field Museum in Chicago. After telling her where the fossilized skeleton of a Dinosaur had been found and how many millions of years ago it lived, she exclaimed, "But how did they know its name?".

The young of *Icthyophthirius multifiliis* attack the skin or gills of fresh-water fish, become embedded in the tissue and form pustules in which they grow to maturity. A heavily infected fish appears as though it were suffering with a type of glorified small-pox; certainly no fish could possible feel as badly as such a specimen looks. When full-grown the ciliates break out of their snug quarters in the skin of the fish, and fall to the bottom where they undergo an orgy of reproduction. One ciliate may divide again and again until as many as 1000 young are generated by a single overworked parent. We, of course, feel sorry for the fish that become parasitized, but we might also weep a little for the young ciliates, since only a few of the most fortunate of them ever succeed in attaching themselves to another fish, the rest starve to death.

Fish are also the favorite hosts of a group of sporozoan protozoa known as Myxosporidia. This type of protozoon is characterized by the production of spores. These spores are usually pear-shaped or spindle-shaped and often less than

1/2500 of an inch long. Nevertheless they are complex in structure, consisting of two valves within which, at the posterior end, are two nuclei and a vacuole, and, at the anterior end, two polar capsules, each containing a coiled and probably hollow polar filament. When such a spore is swallowed by a fish, the digestive juices stimulate the polar capsules to explode, the polar filament is shot like a harpoon into the intestinal wall, and the amebula within, crawls out, pene-

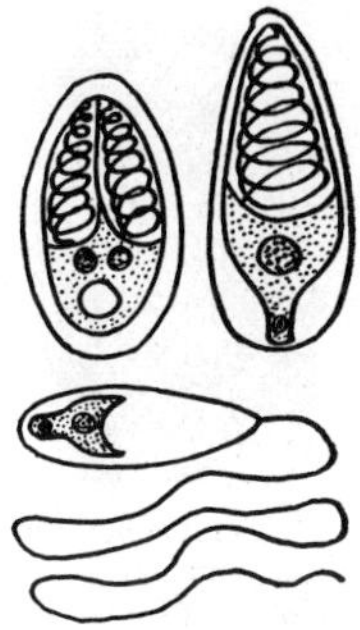

Sporozoan spores, unexploded (above) and exploded (below).

FIG. 36. Those who are afflicted with boils can sympathize with each other although they don't speak the same language.

trates into a blood vessel and is carried to that part of the body in which it is accustomed to live. Here it grows at the expense of the surrounding tissue and seriously interferes with the physiological processes of the fish. Different species of Myxosporidia live in different tissues. When present in sufficient numbers, they are responsible for such diseases as "wormy" halibut, "twist disease" of salmonoid fish, and "boil disease" of certain European species. The bible says, "So went Satan forth from the presence of the Lord, and smote

Job with sore boils from the sole of his foot unto his crown." Its too bad that Job didn't know that fish have boils or he would have had someone to sympathize with.

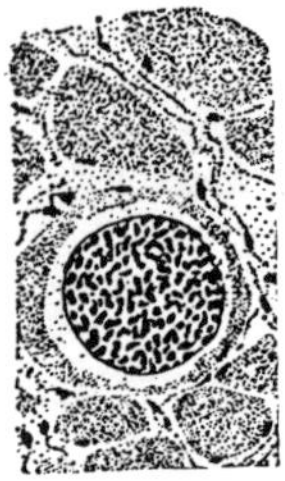

Section of muscle (human) containing a cyst full of sarcosporidian spores.

Spores of sarcocystis

PARASITIC PROTOZOA IN MUSCLE TISSUE

These Myxosporidia may be included among the tissue parasites. Near them in the Class Sporozoa are usually placed another group of tissue parasites, the Sarcosporidia, a name derived from the Greek word *sarco* meaning flesh. Sarcosporidia are spore-bearers that seek out larger game than fish, having been discovered thus far only in lizards, birds and mammals. The infection is best known in mice and sheep but much still remains to be learned about this peculiar type of muscle parasite. It is possible to infect mice in the laboratory by feeding them spores, but no one has yet determined how transmission occurs in nature, although a large amount of work has been carried on with sheep.

After the organisms reach the muscle of the host they form long tubes among the muscle fibers in which are developed thousands of sickle-shaped spores. These spores cannot escape from the body until the host dies, hence it seems probable that infection results from the ingestion of spores with contaminated food or drink. One would expect carnivorous animals to become infected as a result of eating other animals containing spores, but Sarcosporidia have been discovered in only a few carnivores, being much more wide-spread among vegetable-eating species. We don't object when mice die of sarcosporidiosis but are not so happy when this happens to sheep, which often become heavily infected. A toxic substance, known as sarcocystin, has been extracted from sarcosporidia; it is powerful enough to kill small animals such as mice and rabbits.

PASTEUR AND PROTOZOAN PARASITES IN SILKWORMS

The protozoon that made Pasteur famous, or that was made famous by Pasteur, whichever you prefer, is the causative organism of nosema disease, or pébrine, in silkworms. This disease was present in silkworms in other countries as well as in France. It was known in Italy as gattine, a term derived from gattino, meaning kitten, because diseased worms assumed a peculiar attitude with heads raised and claws out like a cat ready to scratch. Sick worms were noted as early as 1845 and by 1850 silkworm raising in France was threatened with utter ruin. No one else seemed able to do anything about it so in 1865 Pasteur was persuaded to undertaken an investigation. He knew practically nothing about silkworm culture, but soon learned what was known and attacked the problem in a scientific manner. Unfortunately two diseases were confused, one was due to microsporidian protozoa; this was given the name pébrine from the word *pébre*, which means pepper, and refers to the spots on the diseased worms resembling grains of pepper. The other was a filterable virus disease called flacherie.

Cells in the stomach wall of an insect suffering from nosema disease. Each dot is a microsporidian parasite.

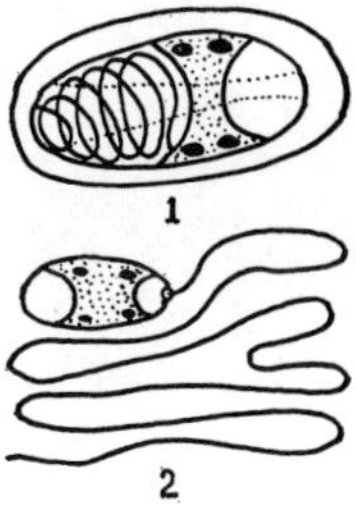

Spores of the nosema parasite of silkworms. 1. Unexploded. 2. Exploded.

In 1849 minute spores, or oval corpuscles as they were called, were noted in sick worms but whether these were responsible for the disease was not determined. Pasteur set out to discover the cause of the disease and methods for its control. His success is now an achievement treasured in the history of biology. The presence of spores was found to be certain evidence of disease and eggs laid by moths free from spores produced healthy worms. The presence of diseased worms that did not contain spores was found to be due to the filterable virus disease, flacherie. Silkworm raising was restored as a result of Pasteur's studies. All the cultivator needed to do was to rear worms from eggs laid by moths free from the spores of microsporidian protozoa. Thus billions of dollars were saved and women still have the pleasure of wearing silk stockings.

HEREDITARY TRANSMISSION OF PROTOZOA

From a biological viewpoint one of the most interesting peculiarities of pébrine disease is the fact that the spores are transmitted through the egg to the next generation. This is known as hereditary transmission, but a distinction should be made between this sort of inheritance, which is somewhat similar to the passing on of property from parent to offspring, and true inheritance which involves the development of characteristics in the young from the actual substance of the parents. Very few cases of hereditary transmission are known among animals.

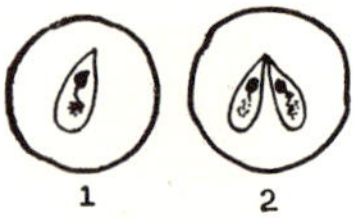

Protozoa of Texas fever in cattle within red blood corpuscles.

An example of hereditary transmission, however, almost as famous as that of silkworm disease, occurs in the life cycle of the protozoan organism responsible for Texas fever in cattle. The causative agent and method of transmission in this

FIG. 37. A young tick full of disease clings to a blade of grass impatiently awaiting the approach of a cow into which it can pump its germs.

case were worked out by members of the United States Department of Agriculture in the earlier years of the gay nineties. The credit is often given entirely to the late Theobald Smith, as has been done by de Kruif (in *Country Gentleman*, 1926, pp. 18, 19, 59), but the more recent review of Hall (in *Journal of Parasitology*, 1935, Vol. 21, pp. 231–243) proves conclusively that, as usual, Smith built upon the work of his predecessors, and was assisted by his colleagues, especially by Salmon, Curtice and Kilborne. In an extensive report published in 1893, Smith and Kilborne reported that Texas fever in cattle was due to a protozoon, that this organism is transmitted from diseased to healthy cattle by the cattle tick, that the guilty organism passes from the mother tick into her eggs, and that the young that hatch from these eggs are diseased and can infect clean cattle. The young diseased ticks react negatively to gravity and climb to the top of blades of grass, where they wait for cattle to pass by. They then transfer themselves to the cattle and immediately inoculate them with protozoan parasites. Soon methods were devised of eradicating ticks by dipping the cattle in an arsenical solution, and ways and means were also devised of freeing pastures from diseased ticks.

PROTOZOA OF HIGHER ANIMALS

The protozoa that live in higher animals are usually separated for convenience into three groups, namely, intestinal, blood-inhabiting and tissue parasites. Those that live in the tissues are all injurious to the host, and hence pathogenic, since they interfere with physiological processes and may actually destroy the cells that surround them or into which they bore their way. Many of the blood-inhabiting species are also pathogenic due to their habit of penetrating red blood corpuscles in which they find a haven safe from the attacks of vigilant phagocytes. But most of the intestinal protozoa appear to be harmless, at least when small numbers are present. They no doubt absorb some of the food that has been digested by the host and hence fall within the ranks of the food robbers, but some of them may even pay for their shelter and transportation by assisting their host to digest certain particularly resistant elements in the food.

When a parasite seems to be harmless to a certain species of host the equilibrium that has been established between host and parasite probably indicates a very long period of association. Nothing is more disadvantageous to a parasite than the death of its host, since almost all species of parasites are thereby buried alive with no means of escape; in the course of organic evolution, those parasites that do not trifle with their luck, but live in harmony with their host, succeed in the struggle for existence, and those that bite the hand that feeds them are doomed to destruction.

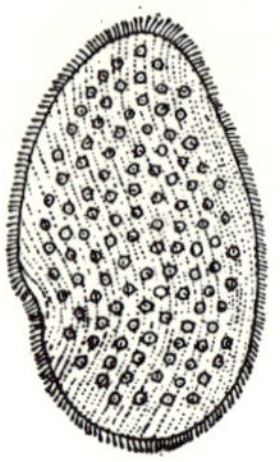

An opalina ciliate

Among the protozoa that have adopted the best policy are many of the ciliates that find a shelter in the digestive tract of vertebrates. If I were a ciliate I should select a warm-blooded vertebrate in which to make my home, but cold seems to have no deterrent influence on protozoa and just as many reside in fish, frogs and reptiles as in warm-blooded birds and mammals.

OPALINA CILIATES OF FROGS

Gems are often found in curious places, but the most vivid imagination would hardly select the intestine of frogs and toads as the home of the gem among ciliated protozoa. Truth is obviously in this case stranger than fiction, since you need only examine with a microscope some of the contents of the intestine of a frog or tadpole and usually numerous specimens of opal ciliates will be your reward. These present an opalescent appearance as they swim leisurely about. Like many parasites, they are modified somewhat by their mode of life; they have no mouth, but absorb their food through the surface of the body, being continually bathed in nutriment digested by the frog. They are literally "in the soup."

In the spring the opal ciliates produce little spherical bodies, called cysts, which pass out of the frog's intestine into the water at just about the time tadpoles hatch from the frog's eggs. The cysts are swallowed by the young tadpoles and hatch in the intestine, thus populating a new generation of frogs.

INTESTINAL CILIATES OF HORSES AND CATTLE

Cattle and horses are among the most abundantly infected of all warm-blooded animals with ciliates. Forty or more different species of ciliates are known to live in the cecum and colon of horses. Here they find plenty of partially digested food that they can filch from their host, and, with nothing to do but eat and reproduce, easily succeed in multiplying beyond all reason. That such swarms of ciliates can live so closely crowded together appears almost miraculous. To them the black hole of Calcutta would seem like home sweet home. Apparently the horse is as unconscious of their existence as most of us are of our internal parasites. Neither the ciliates nor the horse are injured or benefited by the association.

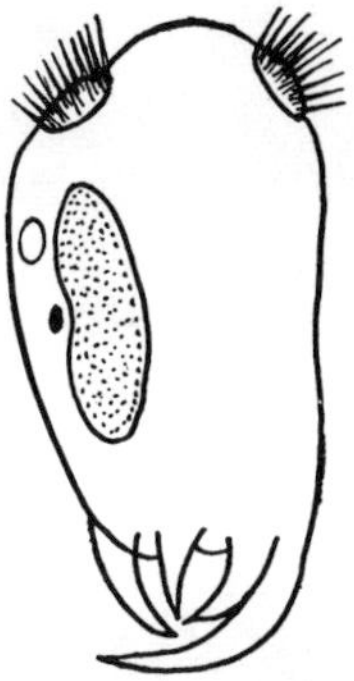

A ciliate that lives in the cecum of the horse.

The rumen, or paunch, and reticulum, or honeycomb, of the stomach of cattle and other ruminants are even more abundantly supplied with ciliated protozoa than are the cecum and colon of horses. Here succulent or dried green plants and grain are mixed with large quantities of saliva and water and maintained at a comfortable, even temperature, thus creating a marvelous culture medium for the activities of protozoa. Life here, however, is not entirely without hazards, since ciliates may be regurgitated from the paunch into the mouth in the animal's cud and in the mouth are subjected to the grinding action of the teeth. Those that escape from this ordeal pass into the honeycomb chamber and may then be carried into the fourth chamber, or true stomach, where they are promptly pickled in its acid contents. The remains of these protozoa are digested, absorbed and assimilated, hence every porterhouse steak we eat is flavored with the essence of ciliated protozoa.

One ingenious investigator has been able to eliminate the protozoa from the stomach of cattle by introducing a solution of copper sulphate through a tube. Animals treated in this way and thus relieved of their protozoa thrive just as well, and seem just as happy, as infected cattle. Evidently in this case the protozoa could easily be dispensed with. Transmission from one animal to another occurs by contamination. When a cud-load of ciliates is regurgitated into the mouth the saliva becomes alive with them. This saliva may be left on partly eaten food or in watering troughs where uninfected cattle may ingest and swallow a sufficient number of the ciliates to set up a new infection.

A ciliate that lives in the stomach of cattle.

PARASITIC FLAGELLATES

In many fields of biology no new worlds are left to conquer. Thus the earth has been diligently searched as with fine-toothed combs for birds and mammals and new species are now added to the list of these large vertebrates with considerable difficulty, except by the splitters who delight in dividing each species into two or more and thus burdening us with

FIG. 38. Millions of ciliates are digested, absorbed and assimilated by cattle daily so that we may have well-flavored steaks to eat.

new scientific names. But protozoologists still have a virgin field and can find new species and new conditions to investigate in even the most common domesticated and wild animals.

Among the intestinal protozoa, probably no genus excels the flagellate *Trichomonas* in numbers of unknown species. Collectors are almost certain to find these flagellates in the intestine of almost every mammal he examines. Most of the trichomonads appear to be harmless but within the past few years two species have been found guilty of serious injuries to cattle and pigeons. The cattle trichomonas invades the genital tract and brings about abortion. The pigeon trichomonas occurs in the crop and may invade various organs, especially the liver. In this case, as with the termites, the parent passes on the infection to 100 per cent of her young when she feeds them on pigeon's milk from her crop.

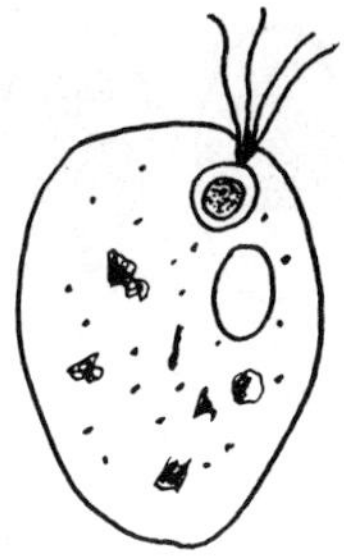

The parasite of blackhead in turkeys.

Another flagellate that is pathogenic to birds, is known as *Histomonas*, and is responsible for blackhead disease in turkeys.

FIG. 39. Picture the consternation of a pilgrim father who anticipates a delicious thanksgiving dinner only to find that all of his turkeys have been destroyed by blackhead disease.

Birds infected with this insidious organism do not feel like gobbling, since they suffer from severe inflammation of the liver and intestine, and in course of time their beautiful red comb turns black. The transmission of this protozoon has puzzled investigators for many years. If the results published recently are correct the protozoa enter the eggs of a parasitic roundworm, *Heterakis*; turkeys swallow these roundworm eggs in contaminated food and when the eggs hatch in the turkey's intestine the protozoa are liberated. Thus a parasitic protozoon depends on hereditary transmission through the egg of a worm for its continued existence. Such a sequence of events certainly constitutes a wonder of nature. Incidentally if these parasites were allowed to spread unchecked we would certainly have to modify our thanksgiving customs.

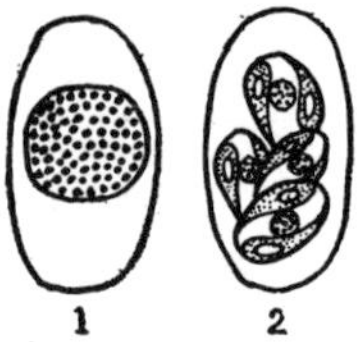

Spores of the coccidium of the rabbit. 1. Undeveloped. 2. Developed.

PARASITIC SPOROZOA

Among the intestinal protozoa are usually included certain sporozoan species known as Coccidia. These are always pathogenic since they penetrate the cells of the intestinal wall, but some species of hosts are more badly damaged than others. Chickens and rabbits are particularly susceptible. The life cycle of this type of protozoa is quite complicated; it resembles in its main features that of the nearly-related malaria parasites which will be described later. The most important stage in the life cycle of a coccidium is the oocyst, which breaks out of the tissues of the intestinal wall into the lumen of the intestine, and is carried out of the body in the excrement of the host. This oocyst is protected by a resistant covering and can withstand conditions outside of the body of the host that would destroy most other protozoa. Young chickens that swallow too many of these oocysts suffer from a severe intestinal disturbance and large numbers of them die. Likewise the mortality of young rabbits is often high as a result of coccidial infection.

BLOOD-INHABITING PROTOZOA—TRYPANOSOMES

Many of the most important human and animal diseases, especially in tropical or semitropical countries, are due to these intestinal protozoa and to another group of protozoan parasites that live in the blood. Among the human diseases due to blood-inhabiting protozoa are malaria, African sleeping sickness, Chagas disease, oriental sore and kala-azar. Malaria organisms occur in mammals and birds as well as in man, and protozoa similar to them are known from reptiles, amphibians and fish. The trypanosomes that are accountable for diseases in man and in domesticated and wild animals were first discovered in the blood of fish in 1841 by Valentin. Two years later flagellates of the same type were reported from frogs (Gruby, 1843). Then in 1878 trypanosomes were encountered by Lewis in the blood of a mammal, the rat. These observations attracted no particular attention, but in 1880 Evans suggested that trypnosomes might be responsible for a disease of horses known as surra, which was later proved to be correct.

Students of pure science usually pay no more attention to pathogenic organisms than to nonpathogenic species, but any protozoon responsible for a disease in an animal of use to man is certain to be thoroughly investigated. Thus Evan's work stimulated a number of microbe hunters to devote their time to the study of trypanosomes. As a result Bruce in 1896 worked out the transmission of a trypanosome disease of horses and cattle in South Africa, the transmitting agent being the tsetse fly. He was the first to demonstrate that a protozoan disease could be transmitted by an insect. The reaction to these studies was mild compared with the furor created by the discovery of a human trypanosome by Dutton in 1902. Since that time scientific periodicals have been deluged with reports on trypanosome diseases in man. We now recognize two types of human sleeping sickness in Africa transmitted by tsetse flies and Chagas disease in parts of South and Central America transmitted by kissing bugs (see Chapter VIII).

Blood-inhabiting protozoa are more restricted in their distribution than are intestinal protozoa because they depend for their transmission on some intermediate host and hence are unable to exist, even where a suitable host occurs, if an intermediate host capable of transmitting them does not live in the same environment. Thus malaria cannot exist in a locality unless certain species of mosquitoes are present to carry the infective organisms from one host to another. Among the lower animals the trypanosomes are the most important blood-inhabiting protozoa. These minute flagellates live in the blood plasm but do not penetrate the red blood corpuscles. Only a few of the 400 or more species of trypanosomes that have been described are known to be pathogenic.

Fortunately animals in this country are relatively free from pathogenic trypanosomes. Several that occur here, however, are of great biological interest. The frog, for example, is infected with the species to which the scientific name *Trypanosoma* was first applied by Gruby in 1843. This species and a species that is often present in the blood of the crimson-spotted newt can be obtained very easily for classroom study. The transmitting agents in the life cycle of these species are leeches. No doubt those of us who have patronized the old swimming hole have on many occasions been pumped full of these trypanosomes by leeches. Fortunately most of the parasites of lower animals, as in this case, are unable to live in human beings.

FIG. 40. The wily trypanosomes allow leeches to inoculate them into a terror-stricken and innocent frog.

A species of trypanosome that occurs in horses, mules and cattle in North America and has been studied extensively in Panama is especially interesting because the vampire bat has been implicated as a transmitting agent. This bat is not the large blood-sucking monster of popular belief but is only about three inches long and laps up blood after making a painless incision. He has the unique distinction of being the only mammal known that transmits a protozoan disease, a very doubtful honor, however, since many of these bats are being captured by scientists who tear them away from their detestable night life and use them as experimental animals in the laboratory.

Certain game animals of Africa may be infected with trypanosomes transmitted by tsetse flies. Others that are pathogenic to cattle and horses are also carried by tsetse flies n Africa. Surra in horses and other animals is transmitted by biting flies such as the horse fly. Another disease due to trypanosomes is mal de caderas: this occurs in horses and mules in South America.

The various modes of transmission displayed by trypanosomes are truly remarkable. Among the transmitting agents, we have mentioned leeches, blood-sucking tsetse flies and kissing bugs, blood-lapping vampire bats, and biting flies. At least one trypanosome disease of horses has no intermediate host; this species is responsible for dourine, or horse syphilis, and is transferred from animal to animal directly during coitus. As though further to insure the maintenance of the race the trypanosomes responsible for African sleeping sickness in man also live in certain wild animals which serve as reservoir hosts, and those responsible for Chagas disease in man make use of the armadillo for a similar purpose. Thus parasites that might kill off all available human beings and consequently annihilate themselves, save the race by adopting other species of hosts in which they are not pathogenic.

STRUGGLE FOR EXISTENCE AMONG ANIMALS

Only those who study plants and animals in the field realize what a frightful struggle is continually in progress both night and day to maintain individual existence and that of the race. And very few of us know how wide spread are diseases among the apparently carefree birds, mammals and other wild animals. Ruskin says "No human being, however great or powerful, was ever so free as a Fish;" but as we sit at the other end of our fishing rod and think of the small chance a fish has of reaching maturity and of the various disgusting diseases he is almost certain to acquire if he does grow up, we are inclined to doubt Ruskin's knowledge of natural history. Similar conditions prevail throughout the animal kingdom. Every species of wild animal that has been examined properly has been found to be infected with one or more species of parasitic protozoa or worms. Bacterial diseases and parasitic insects are also wide spread among wild animals. On a recent collecting trip in the jungles of Panama we shot tiger cats suffering with pneumonia and monkeys afflicted with ulcerated teeth. Many wild animals including monkeys and birds are infected with malaria organisms, fortunately different from those that live in man. However, these and the other types of protozoan parasites that I have discussed in this chapter are of the greatest importance to us, not so much because of their economic significance, but because they are being studied extensively in our laboratories and what we learn from the study of amebas, trypanosomes and malaria in lower animals can be translated more or less directly into human terms. Many of us believe there is no better way in which to spend our lives.

Mother Goose was not a parasitologist, but if she had been, she probably would have modified one of her famous poems so as to read as follows:

Sing a song of parasites,
Of bugs that writhe and squirm,
Flagellates and ciliates
And every sort of germ.

When the host is opened,
The germs are on the spot.
What a dreadful life they lead!
Their lot is not so hot.

Here's little Opalina
In the rectum of the frog,
And old man Balantidium
In the cecum of the hog.

We prefer the blood stream;
It's a cosy little cove.
Wouldn't that be lovely?
I'll tell the world, by jove.

Chapter IV

THE COVERED WAGON MARCHES ON

Adventure lies over the ocean
For pioneers craving to roam;
But nature is swarming with microbes
In Everyman's own home sweet home.

Get wise then to rat and to cockroach;
Be gentle with guineas and geeses;
Inside of their cute little tummies
May flourish a cosmos of species.

THE discovery of a new species of bird or mammal is comparatively difficult but one does not have to leave home in order to find new species of protozoa, since the earthworms, sparrows, crickets, slugs, and other animals that live in our immediate environment are well supplied with parasitic species that have never been seen by any human being. Even domestic animals may be included in this category. Who would suspect the rat or the rabbit of harboring unknown species of protozoa? Yet within recent years three new species of trichomonad flagellates have been described from rats and four new species of coccidians from rabbits. Exploring the digestive tract, blood stream and tissues of any species of animal for parasitic protozoa is therefore a type of pioneering that is available to anyone with an inquisitive mind and the instincts of a naturalist.

According to an old recipe for cooking a hare we are told that we must "first catch a hare." This advice, properly modified, applies equally well to studies in protozoology since we must first learn where to find protozoa before we can determine how they spend their evenings. Fortunately, catching protozoan "hares" is a simple and inexpensive pastime as our experience with giardia flagellates easily demonstrates.

PIONEERING AT HOME WITH GIARDIA

As an example of pioneer work done at home may be cited the study of certain flagellates belonging to the genus *Giardia*. When in 1922 we became interested in these intestinal flagellates the general impression among medical men was that a

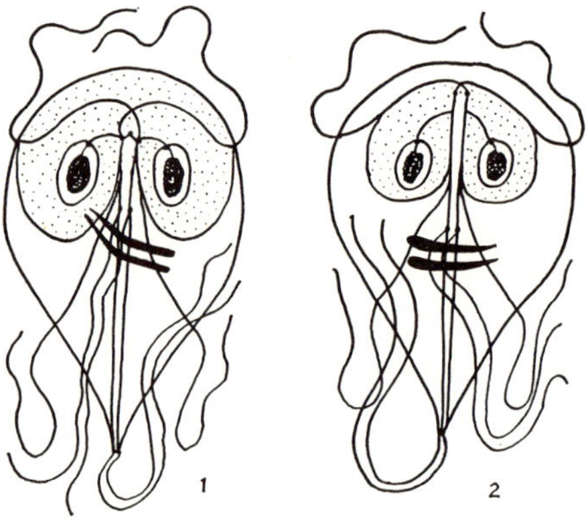

FIG. 41. 1. Giardia from the rabbit, with large sucking disc. 2. Giardia from the civet cat, with small sucking disc.

single species lived in man, rabbits, rats and mice, and that human beings usually became infected as a result of swallowing giardia cysts from rats and mice with which their food or drink had become contaminated. This belief was due to the fact that this group of protozoa had been studied very little by modern cytological methods. It was known to protozoologists that the giardias of field mice differ in shape and size from those of house mice and that their parabasal bodies are elongated and pointed instead of short and oval. Further investigation disclosed similar, but different, characteristics in giardias from rabbits. The rabbit giardias are 50 per cent longer than those from rats, have long, narrow parabasal bodies, and the lateral shields bulge out as though the organisms were suffering from the mumps. The giardias from guinea-pigs look very much like those from rabbits but are consistently only two-thirds as long. Incidentally those who have studied protozoa intensively have found that size is as definite and constant a heritable characteristic of these little animals as it is of larger species such as birds and mammals.

The giardias of the civet cat are noteworthy because of the small size of the sucking disc with which they cling to the intestinal wall. The civet cat is not a pleasant sort of animal—about as desirable an acquaintance as a skunk—and giardias really should not associate with this species. Certain birds are infected with giardias in which the sucking disc is relatively long and the two nuclei within the body lie closer to the lower end of the sucking disc than do those in other species. Recent studies with giardias not only prove our contention that pioneer work with protozoa is still possible, but also furnish data that have an important bearing on the fascinating problem of host-parasite specificity, that is, the distribution of different species of parasites among the various species of available hosts. If we knew more about host-parasite specificity we could explain why pigs and man are both infected with the same species of balantidial ciliates, why a baby that makes a practice of sucking the dog's tail does not come down with canine giardiasis, and why scavenger birds, such as the vultures, are relatively free from parasites.

QUICK AND EASY PIONEERING

Another feature of scientific research that adds to the pleasure of the investigator is the small amount of effort necessary in order to become an expert, provided the field of research is sufficiently limited. Students who have had only an elementary training in biology may undertake a problem in protozoology and within a few months may acquire a more extensive and intensive knowledge of the particular subject he is studying than anyone else on earth. Furthermore, it is a relatively simple matter to become acquainted with all of the species that have been described in certain genera. For example, a resumé of what we know about the genus *Giardia* could be assembled in a few weeks if a good scientific library were available, although a reading knowledge of certain foreign languages, notably German and French, would be necessary. Then an examination of domestic and wild animals for giardias would no doubt reveal species that have not yet been described. Pioneering in protozoology can thus be done at home as well as abroad, and the pioneer is not subjected to such hardships as we usually associate with those who prepare the way for

Fig. 42. The protozoologist who runs away from the gangsters of the microscopic world will live to hunt for microbes another day.

others to follow. In fact, many of us actually earn a comfortable living carrying out explorations in well-equipped laboratories amid pleasant surroundings.

PIONEERING ABROAD

Although a protozoologist who is not able to travel widely in his own country or in foreign countries can find plenty of material to work with at home, many of us have opportunities to make surveys abroad. A certain amount of risk is involved in expeditions to the less civilized regions of the world; methods of locomotion are often primitive and unreliable; sometimes the natives are hostile; and frequently the fugitive protozoa one is trailing turn the tables on the protozoologist becoming the aggressors and conferring upon their enemy such unpleasant and dangerous afflictions as dysentery and chills and fever. However, surprisingly few explorers lose their lives in foreign countries; most of them seem to be killed by automobiles or in airplane accidents after they return home.

KALINGAS AND CROCODILES IN THE PHILIPPINES

On a number of occasions we have had opportunities to examine animals in foreign countries. For example, while in the Philippine Islands we endeavored to determine the character of the protozoan fauna of some of the indigenous birds and mammals and of some of the socalled non-christian tribes. One expedition was devoted to an attempt to obtain crocodiles for study. This trip is described here *in extenso* as an example of the kind of experiences one encounters when he ventures off the beaten path in search of parasites. My companions on this trip were a Chinese student, Mr. H. J. Chu, and a Filipino assistant, Mr. F. Manipol. The School of Hygiene of the University of the Philippines furnished a small automobile for this expedition and a Filipino chauffeur.

We set out with no more definite information than the rumor that we might encounter crocodiles somewhere in the Cagayan Valley. I am sure the animals we were hunting would have shed buckets of crocodile tears if they had known through what hardships we passed trying to capture them. We stopped at the municipal building in every town we passed through and

Manipol consulted with the natives regarding the probable lairs of the wily crocodile. Why any one should wish to obtain nasty little parasites from crocodiles was never understood by the provincial Filipinos, but they were uniformly courteous and did their best to help us. Perhaps they considered us harmless lunatics and thought that by being kind to us they were acquiring merit that would serve them in the hereafter.

By the time we reached Ilagan it was getting dark so we decided to stop there for the night. Ilagan lies near the center of the Cagayan valley. This valley is very flat, largely covered with tall grass and only a small part of it is cultivated. What a boon such an area would be to the Japanese! Not far away on either side is a series of low mountains which in the early evening appeared indistinctly through a purple haze. Any kind of shelter is satisfactory on such a trip as ours, so we were happy to "hole in" at a native hotel. The rooms in such an inn are bare and the partitions are one board thick and do not reach to the ceiling, hence the slightest noise can be heard throughout the entire building and sleep is practically impossible until the last person has gone to bed. Then of course one has to contend with the noises of the night, as well as with nocturnal animals including bedbugs, cockroaches and fleas.

Fleas are the champion leapers of the animal world. They can jump eight inches vertically and thirteen inches horizontally, that is, about a hundred times their own length up into the air. Crickets, grasshoppers, kangaroos and pedestrians trying to escape automobiles while crossing a city street are obviously outclassed. Translated into human terms, an average-sized man with a flea's jumping ability could leap a distance of eight hundred and fifty feet and over a building five hundred and sixty-seven feet high. In spite of this the flea is considered a degenerate animal. To escape flea bites one must be over eight inches from the floor. Fortunately we were very tired and I am among those who are blessed with immunity to bedbug and flea bites.

At Ilagan we were told that crocodiles were plentiful near

the town of Cabatuan which is situated on the banks of the Magac river somewhere near Cauayan. So we drove back to this town and Manipol was soon busily engaged in trying to find horses to take us the twelve kilometers to Cabatuan. There were no horses to be had so we rather rashly decided to leave the car in charge of our chauffeur and walk to Cabatuan. Only one who has walked much in the tropical sun knows what such a trip is like. There was not a particle of shade to

Fig. 43. Would that the United States might enter big and little fleas for the Olympic games.

be had, and we were wearing high boots and carrying heavy packs on our backs. By the time I reached Cabatuan I felt certain those twelve kilometers had lengthened into forty. I was desperately tired and thirsty and felt as though I could never walk again. Manipol was of sturdier stuff, and while Chu and I lay in the shade on the floor of a deserted hut, he set out to discover where the crocodiles were to be found.

By the time I had come to life again, Manipol returned with an encouraging report of his investigations. He had encountered a Filipino named Aricayo who offered to get horses for us and to guide us across the Magac river, and through the jungle to a barrio named Sili located seven kilometers to the west. Here a few Filipinos under Aricayo's supervision were living in the midst of a colony of non-christian natives known as Kalingas. The Kalingas were said to know where crocodiles lived and how to catch them. Aricayo, Manipol and Chu were small in stature and light in weight and thus of the proper type for Filipino horses to carry. I was almost twice as heavy

FIG. 44. A Filipino horse suffers even more than a well-nourished American protozoologist when subjected to a tropical sun.

as any of the others and most decidedly a misfit both from my standpoint and that of the horse.

First we had to cross the Magac river. This was no joke, because, although the river was rather narrow it was very deep and very swift. We secured the services of a boatman with a flat-bottomed boat, slid down the steep twenty-foot bank, jumped into the boat and pulled our horses into the water. Then each of us hung onto a horse's reins with one hand while it swam alongside, and paddled furiously with the other until we reached the opposite shore several hundred feet down stream. Riding through the jungle is always of interest. The temptation to loiter along the way on such a trip as this and pick a few orchids or examine a huge camouflage-marked moth resting on a tree trunk, or watch the activities of the gorgeously-colored birds screaming in the tree tops, is almost irresistible. But time was fleeting and we were forced to push on if we wished to catch crocodiles that night.

Finally we reached the barrio of Sili where Aricayo had cut

FIG. 45. These Kalinga Belles ordinarily wear nothing but a skirt but load themselves with finery when they have their pictures taken.

away enough of the jungle to build a typical Filipino house, and to start a tobacco plantation and had invited some of his friends to join him. Thus the barrio was established. At the time of our visit this consisted of about six Filipino families.

We at once made our wants known to the Kalingas and plans for catching crocodiles were soon under way. Pieces of dry, hard wood about two inches thick and one foot long were sharply pointed at the ends with a machete. A furrow was dug around the center of each and a stout rope fastened in the furrow. Then Aricayo stepped out with a gun; a few minutes later we heard a shot and a howl; and he returned with the bloody carcass of one of the many dogs that were always underfoot. This was cut into quarters and one quarter impaled on each stick. When this was accomplished, we set out on our horses again to a lagoon several kilometers to the west accompanied by half a dozen Kalingas on foot. Here the ropes were tied to trees and the bait was tossed out a few feet from shore.

Crocodiles are the giants among living reptiles. Their teeth are conical and sharp and hence not fitted for chewing. When feeding, they rush at their prey, grasp it in their powerful jaws, where it is firmly held by the teeth, and proceed to swallow it whole. They are particularly fond of dogs. During the day crocodiles lie in the sun on a mud bank, hide in the rushes, or float at the surface with only the eyes and nostrils above the water. Or they sink to the bottom where they may remain for four or five hours without drowning. The vital processes of crocodiles are very slow, hence food is needed only at long intervals. What a lazy life they lead and what a pleasure their housekeeping must be! They don't even hatch their own eggs but cover them with muck, which ferments, thus producing heat sufficient for the development of the young.

Since crocodiles are nocturnal animals nothing was expected to happen until nightfall, so we returned to the barrio. During the night the crocodiles, which were known to live in the lagoon, were expected to swallow the bait provided for them, with the pointed stick parallel to the esophagus. When they tried to swim away the stick was supposed to turn so as to extend crosswise of the esophagus and thus effectively hold the crocodile until we arrived next morning. This method of capture is not pleasant for the crocodiles but most gratifying to the hunter of parasites.

The trail to the lagoon lay through tobacco fields and by the homes of many of the Kalingas. These are simple structures built in trees in a clearing. One or more trees are used as posts on which the floor rests, and the trunks of other small trees are cut the proper length and added as reinforcements. Small logs are fastened horizontally in natural forks in the posts, across which a floor of saplings is laid. This serves as a platform on which much of the daily life is spent, that is, where food is cooked, traps made and siestas enjoyed. A second

FIG. 46. The non-christian Kalingas build their houses in the tree tops. This candid camera shot was made when Mr. and Mrs. Kalinga chanced to be at home.

story with a thatched roof serves as a store house and sleeping chamber. The ladders by which the platform and bedroom are reached are, of course, drawn up at night. It must be rather pleasant to live in this way so high above the ground, and besides it is safe from the attacks of enemies. Just before leaving Manila a report appeared in the daily press of the death of a Filipino who, while sleeping on the floor of his hut, was impaled on a long machete thrust upward from beneath the house.

Not a moment in the Aricayo household was without interest. The house boy, who was really an adult man with a family, came from Ilacano. He was a member of a party of four who

FIG. 47. A typical Filipino house in the country built several feet above the ground so as to be dry during the rainy season. The pig and chickens live under the house.

six months previously had set out to hunt wild pigs. He and two of his companions had an argument with the fourth member of the party whom they eventually speared with fatal results. In order to save his life our house boy turned States evidence, but he was afraid to return to his home and was hiding away among the Kalingas. It seemed rather strange at first to be waited on by a murderer, but no more harmless-looking creature could be imagined. Human life among these natives does not seem nearly as precious as it does to us, and besides, as Hilario quaintly observed, the spearing would never have taken place if the murderers had been sober. Drunkenness to him seemed a satisfactory excuse for any crime. In this case punishment could hardly be considered severe. To be sure, Hilario had "plenty of nothing" but his work was easy, the food was good according to his standards, and life at Sili seemed like one long holiday.

Aricayo's house was not built for much company, but one of the storerooms that was not completely filled with tobacco was placed at our disposal. Here we tried to sleep on a bamboo floor with a bundle of tobacco leaves as a pillow. I discovered that it is bad form in that region to take one's clothes off at night and don pajamas so undressing consisted in taking off our boots. I demanded and got a quantity of muslin with which to make a mosquito-proof tent for protection from malaria-carrying mosquitoes. But sleep was impossible. I was not accustomed to sleeping with my clothes on; my feet were covered with blisters due to the long walk from Cauayan; other parts of my body were sore from riding that Filipino horse to Sili; the odor of tobacco was almost overpowering; Mrs. Aricayo was suffering from consumption and coughed at intervals thoughout the night; her coughing stimulated the dogs to bark and the roosters to crow; and the Filipino family living a hundred yards away were celebrating the first monthly anniversary of the death of one of their children. This celebration lasted all night and was participated in by the entire community. It was characterized by bright lights, plenty of food and drink, loud talking and laughing and, at intervals, weird-sounding chants in a minor key that rose and fell in volume as the emotions of the singers surged upward and died down. We were told that the celebration would be continued for three days and three nights but, thank God, we were subjected to only one-third of it.

I was glad to get up at sunrise and arouse my weary body with some strong coffee. During breakfast a group of Kalingas who had been out all night hunting for deer and wild pigs, returned empty handed but with excellent appetites. When we had finished eating breakfast, we set out for the lagoon to haul in the captured crocodiles. Unfortunately the bait had been removed and not a single one had been captured. Whether the Kalingas, who are very fond of dog meat, were responsible for the loss of our bait or the reptiles had successfully removed it we never learned. One more attempt was made, however, to bag a crocodile. At the suggestion of Aricayo a Kalinga was sent back to fetch a dog. He returned with the most forlorn-looking little puppy I have ever seen.

FIG. 48. Kalingas at Sili. The man on the left has a jungle fowl which he recently caught with the trap he is holding in his hand.

Those of us with guns were then stationed behind trees near the water's edge and the little puppy's ears were tweeked so as to make him cry a bit. This was supposed to attract the crocodiles within gunshot. After half an hour of this naïve performance we gave up and proceeded sadly back to the house.

Preparations were begun after lunch for our return to Cabatuan and to our car at Cauayan. We arrived at Cabatuan just as complete darkness enveloped the World and rain began to fall. Before we left Cabatuan the previous day we had engaged a Filipino who owned a bull cart to take us back to the main road at Cauayan. We, of course, expected to start early in the afternoon and not after dark. At first our jehu refused to venture forth, but when I offered him double the price agreed upon, he proceeded to attach his carabao to its two-wheeled cart, we all piled into the shallow box of the cart, and the carabao was persuaded to begin its slow-motion journey across what looked like a tractless expanse of tall grass.

Riding over a rough trail in a bull cart is not a pleasant experience, especially when it is raining and one has passed through several sleepless nights and a supperless day. A bull cart has no springs, no seats and no cushions hence the continuous shocks to one's cramped legs and inelastic spine soon become agonizing. Add to this the mental stress due to doubt as to the driver's ability to find his way, and you get some idea of how we spent the succeeding four hours. How we kept to the trail is still a mystery to me. The driver had a flash lamp with which he explored the surroundings from time to time, muttering to himself in what were probably swear words in the Tagalog language. No flash lamp could possibly penetrate the gloom of that evening more than a dozen feet. However, we finally saw a light in the distance and after using up four hours to travel twelve kilometers, arrived at Cauayan at 11 P.M. Our car was parked in front of the municipal building, with the chauffeur sleeping in the back seat. The only person in the town that was not in bed and asleep was a policeman. He offered to let us sleep on the floor in one of the cells of the local bastile, but after a hasty conference, we decided to push on to Echague where there was a hotel. This proved to be an unfortunate decision, since, after travelling through a heavy fog for ten miles, in danger of running off the road at any moment or of running into some animal sleeping in the road, we arrived at a stream which had to be crossed on a ferry and the ferry happened to be on the other side with no ferryman in sight.

Very few bridges have been built in the Filipino provinces and streams too deep to drive through are crossed by primitive rafts. The ferryman works from daylight to dark and no amount of shouting and honking could budge him until dawn arrived, hence the chauffeur and I tried to get some sleep in the front seat of the car and Manipol and Chu in the back seat. Our wet clothes added to the discomfort of our cramped positions and myriads of mosquitoes from which we had no protection kept us so busy even dozing was impossible. But there seems to be an end to everything. The ferryman appeared at daybreak, we succeeded in getting facilities for washing our

FIG. 49. Our automobile was ferried across many of the rivers by man power on a raft of logs.

hands and face at Echague and put by a delicious breakfast of ham and eggs, coffee and hot cakes (pasta caliente). The rest of our journey was uneventful.

Only those who have taken such a trip can appreciate properly a bath, clean clothes, and a spring bed. Although we did not capture any crocodiles and returned without any parasites, except those we may have picked up from the places we had slept, from the food we had eaten, or from the mosquitoes that had bitten us, we felt that our visit to the Kalingas well repaid us for the hardships we endured on the trip.

RACIAL SUSCEPTIBILITY AND RESISTANCE

One of the problems that has troubled protozoologists for many years is that of racial susceptibility and resistance to infection. Intestinal protozoa have been reported from human beings throughout the world wherever examinations have been made. Likewise there is little or no specific racial differences with respect to infections with blood-inhabiting species such as malaria parasites, the trypanosomes of sleeping sickness and the flagellates of kala-azar. Nevertheless, when foreigners are compared with natives differences in both infectiveness and in the severity of infections appear to exist. These may be due to various factors such as differences in diet, in chances of becoming contaminated, in immunity that might have been built up during childhood, etc. Thus any opportunity to survey a race that does not ordinarily mix with other races is eagerly grasped by a protozoologist.

Among the most interesting inhabitants of the Phillippine Islands are the pigmies who are known there as Negritos. These little people are negroid in type. They are supposed to have occupied the entire country at one time but have been pushed back into the less accessible mountain regions by the larger and more aggressive Filipinos. To us they were of particular interest since they do not ordinarily intermarry with Filipinos and hence represent a pure race of the genus Homo.

PROTOZOA OF PHILIPPINE PIGMIES

An opportunity to examine a few Negritos was offered by a former private in the American army of occupation who had remained in the Philippines, had married a Filipino woman and had established himself near the naval station at Olongopo. The mountains nearby were inhabited by Negritos. It might have been dangerous to try to visit them in the mountains, since they still use poisoned arrows, but the American and his wife often aided certain members of the tribe with gifts of food and could arrange to have them come in for examination.

We took a government boat across Manila bay and around to

the naval station at Olongopo and then an automobile to the American's home in the country several miles beyond the reservation. On the following morning a dozen Negritos arrived. You can imagine how difficult it was to explain to them what we wanted and why we wanted it, but in course of time we were able to obtain thirteen samples at a cost of ten centavos each.

FIG. 50. Those who hunt for protozoa among the pigmies must beware of poisoned arrows.

Every one of the little men and women hesitated before accepting a container but the ten-centavo lure proved too much for them and in the end they agreed to the proposition. When they brought the containers back they wore an expression that could only be called sheepish, and were greeted with a rather ribald horse-laugh by their friends. We wondered what their state of mind was as we set up our microscopes and proceeded to make our examinations.

No one seems to have attempted to determine the number of

investigations that give the expected results as compared with those that do not. About the only satisfaction the scientist has is that once in a while the results are far beyond expectations and the accompanying thrill is ample compensation for previous disappointments. Anyway we are searching for the

FIG. 51. A group of Filipino pigmies, called negritos, under a typical shelter. The only woman in the party is just behind the earthen-ware pot.

truth hence the results are good regardless of what they are. In this case the truth was that these Negritos were heavily infected with the same types of intestinal protozoa that have been reported from races of men in various other parts of the world.

PROTOZOA OF RIO CHICO INDIANS

While working at the Gorgas Memorial Laboratory several years ago we learned of a tribe of Indians in the Darien region of Panama near the Columbian border who practiced a curious method of sanitation. Instead of polluting the soil as so many primitive people do, these Indians are accustomed to deposit their excrement directly into the water of the streams on whose

FIG. 52. A group of Rio Chico Indians in a dugout canoe on the Chucumique River in Darien Province. Panama.

banks they live. Such a habit might so completely dispose of intestinal parasites that are ordinarily spread by contamination as to bring about the elimination of this type of protozoa entirely. Arrangements were soon made to visit these Indians.

We embarked one afternoon on a dirty little oil-burning boat which crossed the Bay of Panama during the night and entered the Tuira river the following morning. All the next day we travelled up this river, stopping at La Palma and Chepigana on the way, and as twilight approached we tied up to the bank at the little village of El Real. Here we remained over night.

The boat continued its voyage the next morning by way of the Chucumique river to the little settlement of Yavisa. The Alcalde (Mayor) of El Real accompanied us to Yavisa and with the aid of his authority we secured Indian boatmen who poled us up the Rio Chico river for a visit to the Rio Chico Indians. These Indians are among the most primitive we have ever seen. After confirming the rumor regarding their method of fecal disposal we arranged for the delivery of samples to the boat the following morning. Samples from only twenty Indians were secured but these were sufficient to give us the information we desired. Contrary to our hypothesis, infection with both protozoa and worms was relatively high. We are not certain how to explain this condition but suggest that river water that has been contaminated is used for drinking purposes by these natives and that infective cysts are frequently swallowed.

PROTOZOA OF ESKIMOS

One race that should be surveyed for intestinal protozoa is the Eskimos. As noted in Chapter VII a carnivorous diet creates a condition in the intestine unfavorable for protozoa. Since the Eskimos live largely on animals they should be relatively free from these parasites. We hope sometime to have an opportunity to test this theory, and are willing to predict that none, or only a few, protozoa will be found.

CUSTOMS THAT PREVENT INFECTION

Several of the non-christian tribes in the Philippines live under conditions that are of considerable interest to parasitologists. A favorite trip from Manila is to the mountain resort Baugio and from there to Bontoc in the Mountain Province of Luzon. Here live the Bontoc Igorots. These non-christians build their huts on the ground instead of on posts as the Filipinos do. Customarily each family owns at least one pig. The pig pen is below the surface, that is, a trench perhaps six feet wide, ten feet long and three feet deep is dug at the back of the hut, with a small part of the excavation extending under the hut. In this cavity the pigs are kept and into it are thrown any waste materials; the latter includes the dejections of the human inhabitants. Consequently there is no soil pollution among the Bontoc Igorots and chances of contamination with intestinal parasites are accordingly slight and the parasite burden is much less than it would be if pigs were not maintained in this way.

Another custom characteristic of the Bontoc Igorots is that of cooking their meals on the floor in the center of the hut, no effort being made to let out the smoke even through a hole in the roof. One effect of this custom is that the smoke keeps mosquitoes out of the house and prevents the transmission of any malaria parasites they may be infected with. No health officer could offer a more effective method of preventing the spread of malaria.

The origin of customs such as these may not have involved any attempts to prevent infection with parasitic organisms, but these practices are certainly effective. In several countries, notably in China, hot tea is ordinarily drunk in place of water. China is one of the countries that conserves human manure and uses it extensively for fertilizer. Water has been shown to be one of the principal media for the distribution of protozoan cysts and no chemical treatment has been discovered that will destroy these cysts without rendering the water unfit to drink. However, heating water to a temperature of about 115°F. kills any cysts that may be present, hence, although the water used to make tea may be contaminated, the infective protozoa present are non-viable.

Those who travel or reside in China, as well as in many other countries where sanitation is primitive, are advised to be careful not only of the water they drink, but also to guard against organisms that may cling to fresh vegetables. Investigators working at the Peiping Union Medical College have found that the immersion of fresh vegetables in water at 80°C. for 10 seconds will kill any protozoan cysts that might be present.

BALANTIDIAL CILIATES IN PHILIPPINE WILD PIGS

One of the projects that interested us while in the Philippines was the problem of balantidial infections in pigs and man. So far as we know the large intestine of both man and pigs is infected with the same species of balantidial ciliates. We found domesticated pigs that were examined at an abbatoir in Manila to be infected, but we wished to know whether wild pigs were likewise infected. Several attempts to obtain wild pigs by night hunting in regions where they were known to roam proved unsuccessful, but we finally made arrangements with the government commissioner who had charge of the Manila water supply to invade territory that was ordinarily closed to hunters. Our Filipino assistant called for voluntary helpers and when we started out for the battle field our party consisted of nine men with guns and eight beaters with five dogs. The gunmen were stationed at strategic points and the beaters surrounded a considerable area containing small trees and high grass. Only one pig was bagged on this occasion, so each hunter received only a small quantity of pork for his trouble, but this pig was well-infected with balantidial ciliates, which made our efforts worth while. Microscopic examination revealed no characteristics by means of which these balantidia could be distinguished definitely from those of domestic pigs; so far as we could discover they were no wilder than those in the pigs that are kept by many Filipinos as household pets.

PROTOZOA OF FILIPINO BATS

Pigs is pigs and perhaps fortunately for all of us, pigs can't fly; but certain mammalian relatives of the pig antedated "Darius Green and his flying machine" by several million years. How inappropriate is the ugly word "bat" for such twittery, fluttery little chiroptera. Within a few miles of Manila are a number of caves inhabited by small bats and in various localities during the month of October groves of lanzone trees are visited nightly by droves of enormous fruit bats known as flying foxes. Blood-inhabiting protozoa, such as malaria parasites and trypanosome flagellates, have been reported from bats but practically no intestinal protozoa. Whether this was due to the failure of protozoologists to make proper examinations of these animals or to the actual absence of the protozoa we set out to ascertain. Half a dozen bat caves were visited, including one near the famous Taal volcano, and several caves were entered which were free from bats but contained edible bird's nests built by swifts. Bats were shot either within the caves or as they flew out at our approach, but no intestinal protozoa were found in those that live on insects.

Large fruit bats also seemed to be free from intestinal protozoa. We spent several moonlight nights among the lanzone trees and succeeded in securing a few specimens as they flitted about on noiseless wings. The owners had hung lanterns among the branches which were shunned by the bats, so that hunting them was rather difficult.

Why bats are free from intestinal protoza we don't know. Perhaps the animal character of their diet renders the contents of their intestine unfavorable for protozoa as is known to be true in certain other animals (see Chapter VII), but this hypothesis could not be valid for the fruit-eating bats. Another suggestion that occurred to us is that when not on the wing, bats hang from the roof of a cave or from the limb of a tree and their dejecta are deposited on the ground under them. Since they capture their food while in flight, they do not move about on the ground and hence neither their food nor their bodies become contaminated with protozoan cysts. However, the insects that transmit blood-inhabiting protozoa, such as mosquitoes and bugs, have free access to bats and hence these animals are frequently infected with malaria parasites and trypanosomes.

PROTOZOA OF PHILIPPINE ANIMALS AND PLANTS

Our studies of the protozoa of wild animals while in the Philippines necessitated frequent excursions into the mountains of Luzon and to some of the other islands. Several rare species of animals occur in certain of the islands. For example, the flying lemur, *Cynocepalus volans,* was known to inhabit the island of Bohol, and a species of primitive primate known as tarsiers had also been reported from this island. An expedition was accordingly equipped and proceeded to Bohol and fortunately brought back living specimens of both these types. Very few naturalists ever have an opportunity to see either of these animals in their native haunts and probably neither of

FIG. 53. Tarsier

them had ever been examined for protozoa before this time. Protozoa were obtained from them but their exact status has not yet been determined.

Flying lemurs are about the size of a cat. They do not really fly, but, like the flying squirrel, sail from one tree to another on their outstretched "wings" for distances of two hundred feet or more. The skin is extended on all sides, from the head to the fore legs, from the fore legs to the hind legs, and from the hind legs to the tail, so that when the legs are spread out, a large surface is provided that supports the animal like a parachute as it glides through the air.

The tarsier is about the size of a rat, but looks as though it had been constructed from what was left over after the other mammals had been created. The head is round; the eyes are extremely large and owl-like; the ears are very large; the tail is very long and ends in a tuft of thin long hair; the hind legs are much longer than the fore legs and built for leaping; the fingers and toes are long and all bear nails except the second and third toes, which have claws; and the tarsal part of the hands and feet is extraordinarily long, hence the name Tarsier. He is a nocturnal animal, coming out at night to search for insects and lizards, and moving through the trees leaping from branch to branch, and clinging to the bark with disklike pads on the ends of the fingers and toes, much as a tree frog does. Like a squirrel, he holds his food between his hands while eating it.

Among the other interesting protozoa recorded from Philippine animals were a new species of giardia flagellate from the civet cat, large numbers of ciliates from the stomach of the carabao, or water buffalo, giardias from the Indian cattle egret, coccidians from nine species of wild birds, malaria parasites from two species of wild birds, several species of intestinal protozoa from the python, iguana and small lizards, a number of species from frogs, flies, mosquitoes, cockroaches, and snails, and a species of flagellate from the sap (latex) of a plant, *Euphorbia hirta.*

LABORATORY INVESTIGATIONS AT HOME

The discovery and naming of new species is only a minor phase of the science of protozoology, although an important one, since we must know with what organisms we are working before we can become their biographers. After we are able to identify a species we next attempt to work out its life cycle. The general outlines of these life cycles can usually be determined without much difficulty on the basis of our knowledge of those already established, but in a few cases the intermediate hosts are not known, although many competent biologists have spent years endeavoring to discover them. For example, the manner in which the flagellates of kala-azar are carried from one human being to another is still in doubt.

Such surveys as that just described in the Philippines occupy only a small part of a protozoologist's time, and often the investigations carried on in the laboratory at home are equally exciting. Many different projects might be cited, but our experiences in infecting chicks with protozoa from other animals, and in determining the relations between protozoa and several types of plants will suffice as illustrations.

CHICKS AS LIVING TEST TUBES

A few years ago we were able to demonstrate that certain parasitic protozoa too few in number to be found by ordinary

FIG. 54. Consider the fate of the downy chick that serves first as a living test tube and then decorates the festive board on Sunday.

microsopic examination, would multiply so rapidly in culture fluid in a test tube that they could be detected with ease after two or three days. The method used was to collect on the end of a toothpick a small amount of the material to be tested, drop this into a test tube containing culture fluid, and incubate it for two or three days at 37°C. The culture medium is very easily made up. All one has to do is to add $\frac{1}{6}$ of a pound of common salt to ten quarts of water and keep this as a reservoir supply. Then when cultures are made shake up 100 cc. of this with $\frac{1}{2}$ gram of dried blood serum which can be purchased for a few cents. About ten cc. of this is then put into each test tube.

Recently, during certain of our experiments, we discovered that chicks when a few days old answer the purpose of living test tubes. Chicks are particularly favorable for experimental work because they are inexpensive, easy to maintain in the laboratory, and are free from parasites when they hatch from the egg. Furthermore, when the experiment is completed the chicks have usually increased to an edible size and thus the evidence may be used to eke out the notoriously poor salaries paid to professors of protozoology.

Most of the intestinal parasites of birds live in the cecum. Chickens possess two ceca; these are small blind tubes several inches long that open into the intestine a short distance from the posterior end. Material containing protozoa that is injected into the crop of the chick, or into the rectum, through a thin rubber tube, reaches the ceca and may be found in them within a few minutes. Here a surprising number of species will multiply very rapidly and soon become so numerous that they can be found without difficulty.

At first we killed the chicks in order to obtain cecal material for examination, but later we learned to distinguish between the cecal contents and intestinal contents deposited by them. The former were light-brown in color and of uniform and rather liquid consistency. Chicks that are placed on a paper towel and confined under a large glass bell-jar will usually deposit cecal material within half an hour. Samples from the intestine of a number of animals that might contain protozoa were injected into chicks. The hosts from which these samples were taken included man, chimpanzee, guinea-pig, rat, prairie-dog, pig, and fowls and other birds; and the protozoa successfully grown in the chicks comprised trichomonad and chilomastix flagellates and several species of amebas.

Of particular interest was the discovery of several new species of protozoa that could not be found in the samples injected into the chick but that multiplied sufficiently in the cecum to be detectable. For example, cecal material from guinea-fowls, that appeared to contain only one species of trichomonad, gave rise in the chicks to three species of trichomonads, two other species of flagellates, and three species of amebas; several of these were new to science. Cecal matter from other birds gave similar results. No one has yet carried out any extensive survey of human protozoa in which chicks were used as living test tubes, but this technique should prove of real value since in our laboratory experiments we maintained human trichomonads in chicks for a period of 141 days, and human amebas for several days.

PLANTS THAT CAPTURE PROTOZOA

Bladderworts. One of the most peculiar relations between different types of organisms is the capture and digestion of

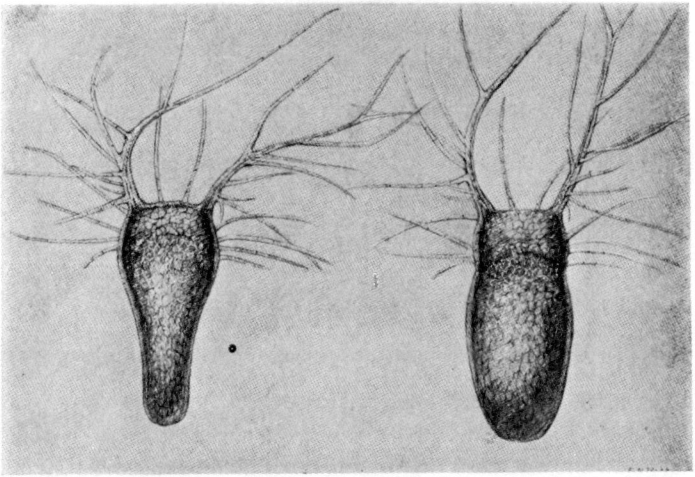

FIG. 55. The bladderwort in action. The bladder on the left is all "set"; that on the right has expanded and captured a poor little protozoon.

minute animals by "insectivorous" plants. The sundew is a favorite example, but lives on land, and limits its hunting to terrestrial animals, hence protozoa are probably never numbered among its prey. Pitcher plants also live on land, but protozoa are carried into their pitcher-like traps by other animals and some of these protozoa are able to live and multiply there successfully. But the most effective plant structures for the capture of protozoa are the bladders of the bladderwort (*Utricularia*). These bladders are located beneath the surface of the water, and, when large numbers are present, may prove to be a real menace to the smaller forms of animal life that swim about in fresh-water ponds. Experiments have demonstrated that the animals found inside of the bladders are not invaders, but have actually been taken prisoner by the plant. Darwin in 1875 attempted to work out the relations between the blad-

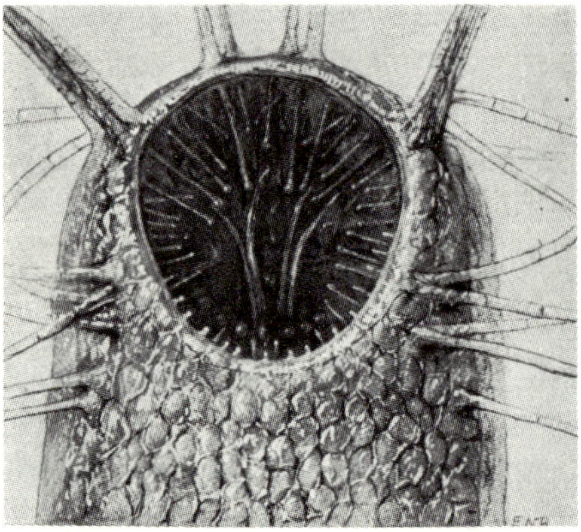

FIG. 56. Looking a bladderworth in the mouth. When the hairs are touched the mouth opens and the bladder sucks in any small organisms that chance to be in the water near by.

ders and the enclosed organisms but failed to discover the way in which they were taken in.

The manner in which the bladderwort captures its prey is as follows. The bladder when "set" has compressed walls as shown in the accompanying drawing. Any small animal that chances to swim into the vestibule of the bladder may come in contact with the various bristles and glandular hairs located on the valve that closes the entrance to the bladder or on the sides of the vestibule. In some way the movements of the animal stimulate the valve to open; then the side walls expand; water is sucked into the bladder with a rush and the prey is carried in with it. This all happens very quickly indeed, in fact, so rapidly that only a chance observation revealed to us the amazing effectiveness of the mechanism.

After a bladder has expanded it immediately begins to set its trap for another catch; 88 per cent of its contents are in some unknown way expelled; and the walls again become compressed in the surprisingly short period of twenty minutes. How effective these bladders are is indicated by the enormous numbers of organisms contained in them. We counted the bladders attached to a bladderwort stem seven feet long and recorded 13,860 of them. The organisms contained in these bladders were mostly minute crustaceans (water fleas), the larvae of insects, rotifers (wheel animalcules) and protozoa. From six to twenty-two crustaceans were counted in each bladder, the average being twelve. This seven foot section of the plant therefore had captured about 150,000 crustaceans, besides mosquito and other insect larvae, and many rotifers and protozoa. Most of these die and disintegrate within a few hours, hence this number probably represents only a single day's hunting.

Protozoa were often found swimming about in the bladders; these were principally of the euglena type although minute ciliates and amebas were encountered. When euglenas were injected into a bladder in the laboratory by placing a collection of them in the vestibule and then springing the trap with a needle, they were not killed but grew and multiplied in the bladder liquid. Paramecia, however, die in about an hour and distintegrate inside of the bladder within several hours after capture. Evidently all of the protozoa, as well as the larger organisms, are captives and not intruders; we have actually seen bladders in nature capture a number of species of protozoa, including euglenas and paramecia. Apparently digestive fluids are secreted into the bladders that kill and dissolve certain of the organisms captured; and this digested material is used as nutriment by the plant.

PROTOZOA OF PITCHER PLANTS

Pitcher plants also capture small animals, principally insects, which they are supposed to digest and use for food. Several species of protozoa have been reported from the pitcher liquid of various species of pitcher plants, where they appear to thrive in the same medium that is suspected of digesting larger organisms. We have found specimens belonging to three of

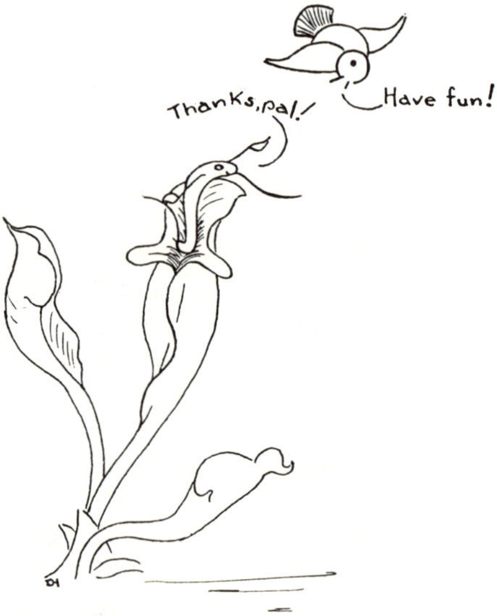

FIG. 57. Protozoa are grateful to the birds that transport them to the pitcher plant.

the four classes of protozoa living in pitcher plants, namely, amebas, flagellates and ciliates. We have also introduced paramecia and other species into the pitchers of healthy plants both in the laboratory and in the field. Paramecia were recovered from these pitchers daily for a number of days and not only thrived but also multiplied in the pitcher liquid. The suggestion has been made that some of the protozoa reported from pitcher plants may be species that are found nowhere else, but the difficulties of transmission from one plant to another seem to rule out this hypothesis. A better explanation is that free-living protozoa are carried into the pitchers on the bodies of insects or birds.

PIONEERING WITH PROTOZOA

I suppose all investigators are pioneers or they would not spend their lives exploring scientific borderlands, but those who study protozoa seem to me to be particularly fortunate. In

FIG. 58. The modern pioneer among the protozoologists carries a microscope instead of a gun and his beautiful assistant is burdened with a calculating machine.

the first place, there is a certain fascination simply in looking at objects under a microscope that cannot be seen with the naked eye, especially when these objects are alive and moving, beautiful in shape and color, and previously unknown. Observing a protozoon that has never been seen by any other human being always gives rise to a pleasant sensation. Protozoa are absurdly easy to "catch," the necessary tools for their study are few and inexpensive, and the space required for laboratory purposes comparatively small. The writer, therefore, does not hesitate to recommend pioneering with protozoa as either a vocation or an avocation, particularly to those who have become a bit tired of stamp collecting or raising tropical fish.

PIONEERING IN THE FUTURE

One of the interesting features of science is the apparently inexhaustible supply of problems waiting to be solved. Furthermore, the solution of one problem usually suggests others, which, like the "little fleas", continue on *ad infinitum*. There is thus plenty of pioneering to be done in the future. Protozoa should be studied in lower animals, especially monkeys, as well as in men. When our ship comes in we hope to study the gibbons of Siam, the orang utans of Borneo, the aye-ayes and lemurs of Madagascar, the tarsiers of Java, the marmosets of Tropical America, the baboons of Arabia, and the chimpanzees and gorillas of Africa,—rather a large order, and also rather improbable of accomplishment, but nevertheless a worthy star for a protozoologist to hitch his wagon to.

Chapter V

ADAM HAD 'EM

A TEACHER of physiology once received a note from an anxious mother which read, "Don't tell Mary about her insides. It ain't nice." Anyone who feels this way about the human body should carefully shun the subject of parasitology, since parasites may occur in every organ and

FIG. 59. Suppose you were singing in your bath and your body suddenly disappeared leaving behind only your parasites, you would become nought but a ghostlike and voiceless shadow.

tissue from bald spot to broken arches. In fact if the body of a person well-parasitized with protozoa were to disappear entirely, leaving behind only the parasites, a perfect outline of the human form would remain and even the features would be so clearly discernible that individuals could be recognized.

If this were to happen some morning while you were singing in your bath you would suddenly become voiceless, since no one has ever heard a parasite make a noise.

It gives one rather a shock to learn that human beings may be infected by as many as twenty-five different species of protozoan parasites, and that, since these parasites exhibit no antagonism toward one another, one person may be infected at one time with the entire job lot. Some people appear to be more susceptible than others, and either do not become parasitized, or if they do, are so tough that the parasites soon give up the struggle,—a condition analogous to that of the rattlesnake that bit Bill Jones and then went away and died.

MAN A FIT ENVIRONMENT FOR PROTOZOA

Man must present a very fit environment indeed to have become the host of such a large number of unwelcome guests. Some of the conditions favorable for the parasites involve temperature, moisture and food. Protozoa are not usually classed along with fish and whales as aquatic animals, nevertheless they cannot live where it is dry but must have moisture in order to carry on their various activities. Since over two-thirds of the weight of the human body consists of water, we present to protozoa a common carrier well suited for their aquatic sports. The human body is equally favorable as regards temperature, since, unless we get too romantic or have cold feet, our internal micro-climate offers uniform warmth without any coal bill or furnace to tend. To be sure, the sun doesn't shine within us and our parasites must live in eternal darkness, but they have no eyes to see with anyway and hence do not miss the light.

The most important environmental factor that we offer our protozoan parasites, however, is probably a constant food supply. Think how contented the protozoa must be that live in our highly nutritious blood stream, or in the intestine after a steak dinner has been well digested. Furthermore, there is no danger of ennui since these media change from time to time, thereby introducing considerable variety. As I hope to show later too much meat is not favorable for intestinal protozoa, hence this type of parasite should select a vegetarian for a host, and those blood-inhabiting protozoa

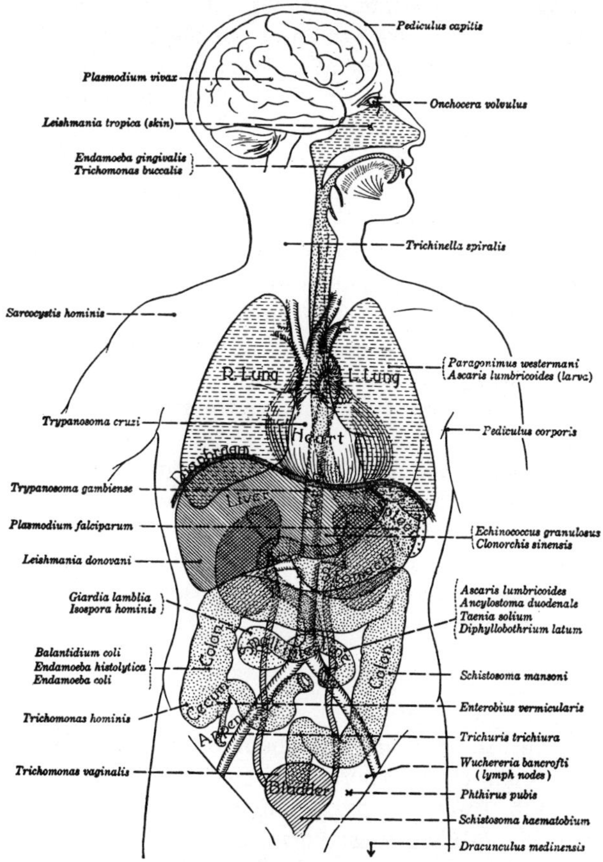

FIG. 60. This diagram indicates where many of the protozoa (left) and worms and insects (right) are localized in the human body.

that have a sweet tooth are advised to pick out a diabetic in whom to live.

Another feature of the human body that favors the parasite

is the absence of gate keepers to prevent the entrance and exit of the protozoa. All human infections with protozoa can easily be prevented by means of a suitable bodyguard. In every case we are responsible for our own infection. We permit the passive parasite to enter our digestive tract in contaminated food or drink, or allow mosquitoes, tsetse flies and other intermediate hosts to squirt us full of such noxious germs as those of malaria and sleeping sickness. Even such a harmless pastime as kissing furnishes an avenue of infection that is about as perfect as one can imagine from the standpoint of both the parasites and the hosts involved.

Few of us who live in northern climes are free from protozoan parasites, but we are usually infected with a lesser number than are the inhabitants of the sunny south. Apparently also, those who live in cities are less often infected than those who live in the country, a situation that is due not so much to the greater sophistication of the city slickers as it is to the lack of sanitation in the wide open spaces. In tropical and semitropical regions, where it is moist and warm throughout most of the year and where life is so easy it seems hardly worth while to try to keep clean, conditions for the transmission of human protozoa reach their greatest effectiveness and a native free from one or more parasitic protozoa would be worth exhibiting as a unique phenomenon.

ORAL AMEBAS

All four of the classes of protozoa are represented among human parasites. We may call them for convenience, amebas, flagellates, ciliates and sporozoans. These organisms can be separated into three groups, as already noted, and designated intestinal, blood-inhabiting and tissue protozoa. Suppose we pass through the digestive tract of one of our friends and see what a strange assortment of wild animals we may encounter. Our expedition is halted almost before we get on board by two species of protozoa hardy enough to withstand conditions in the human mouth. Amebas of the type shown here are present in the mouths of over 50 per cent of the general population. How they get there I can leave to your imagination. Certainly every political candidate should be examined before he begins his campaign so that the babies in his district can be properly safeguarded against his kisses.

These amebas hide in the tartar of the teeth and in the materia alba around them. They feed on leucocytes and bacteria and apparently do no harm. At one time they were

Ameba of the human mouth

FIG. 61. Fortunately for young love, the protozoa that live in the human mouth are harmless.

accused of causing pyorrhea but were later acquitted by the "nine old men" of science. This should be good news, especially in the spring of the year, since our amorous behavior need not be modified for fear of becoming infected with oral amebas. Making your dentist happy by having all your teeth removed has proved to be of no benefit in attempts to discourage the organism, since store teeth seem to be as attractive to the amebas as those provided by nature.

These amebas must certainly be tough babies to withstand the variations in their environment resulting from our outlandish food habits. We may begin our dinner with one or more cocktails containing 20 per cent alcohol, and follow these with highly spiced hors-d'oeuvres. If the chef is good, the soup is almost boiling hot and a drink of American ice water becomes necessary for oral comfort. The fish is well flavored with lemon juice and the meat course includes plenty of protein. Carbohydrate abounds in the potatoes, asparagin in the asparagus, and carotin in the carrots. Occasionally samples of sherbet furnish the amebas with an icy shower bath during the intermissions; and the salad contributes vitamins, a dressing that reeks with oil and vinegar, and a touch of garlic. Perhaps the greatest danger is the high sugar content of the vanilla ice cream with chocolate sauce which raises the osmotic pressure until the amebas become painfully contracted. But there is more to come, including strong coffee, sweets containing peppermint, a liqueur potent with alcohol, and a chesterfield cigarette, satisfying though mild. And while the host is enjoying his dinner the amebas ooze their way hither and yon among the champing teeth trying to escape the pile-driver effect of mastication. Those that are not crushed to death or killed by overeating are in constant danger of being swallowed. This is fatal since experiments have demonstrated that these amebas cannot live in the gastric juice of the stomach. After all, who wants to be an oral ameba?

One peculiarity in the life cycle of the oral ameba is the absence of a cyst stage. Cysts are resistant bodies that represent a phase in the life cycle of most of the parasitic protozoa and are ordinarily necessary for the successful transmission of viable organisms from one host to another. Nature, is a realist however, and seems to understand that oral amebas are transmitted from one host to another by contact and that consequently no resistant cyst stage is necessary for the maintenance of the race.

ORAL FLAGELLATES

Trichomonad of the human mouth

The flagellates of the human mouth are of the trichomonad type. They appear to be beautiful, shimmery little creatures when viewed with a microscope as they move slowly across the field in a zigzag path like an inebriated man searching for a lamppost to lean against. These flagellates are probably less resistant than the amebas since not so many of us are infected by them. They seem to be harmless and are rather to be pitied than condemned. Through the center of the body of each trichomonas passes a slender, flexible rod that serves the purpose of a backbone. At the forward end are four whip-like flagella that beat backward thus drawing the body forward, and along one side the protoplasm is drawn out into a thin membrane; this is longer than the body and hence thrown up into folds. Along the outer edge of this membrane is an active flagellum. This undulating membrane seems to serve as a locomotor organ, and, since no such membrane usually occurs in free-living protozoa, it is supposed to have evolved in answer to the necessity of making progress through the highly viscid medium in the digestive tract. Near the forward end at one side is a comma-shaped orifice into which the flagella may drive solid particles of food. It is believed, however, that most of the nutrition of trichomonads is absorbed through the surface of the body.

Although one must live a strenuous life to exist in the human mouth, the human stomach is apparently an even more difficult habitat for protozoa, since no species are known to make this their home. As noted in Chapter III, the stomach in certain of the lower animals, such as cattle, is a favorite habitat for various types of protozoa. Why the stomach of cattle is infected with protozoa and that of human beings not infected is a problem that involves conditions to which I will refer later.

GIARDIA FLAGELLATES OF THE SMALL INTESTINE

We again encounter parasitic protozoa after we pass through the sphincter muscles that regulate the transit of material from the stomach into the small intestine. Here we enter a tube about an inch in diameter and about twenty-three feet long. This should afford plenty of space to live in and plenty of predigested food to be absorbed and one would expect to find the protozoa flourishing in such an environment. But even California suffers from earthquakes and Florida is visited by hurricanes. Unfortunately the protozoa that live in the small intestine must withstand the force of peristalsis which tends to force them down and out of this section of the diges-

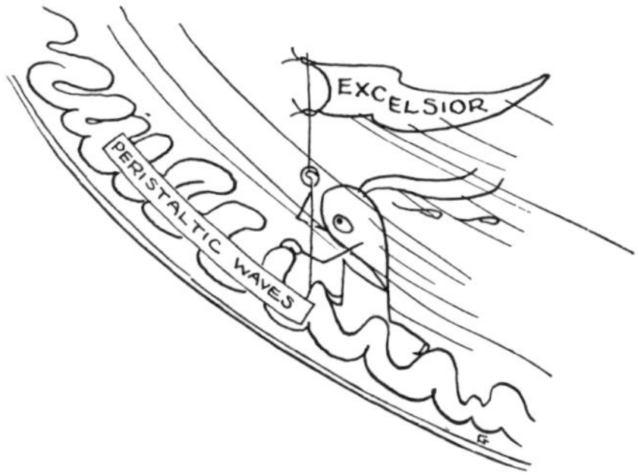

FIG. 62. Life on the peristaltic waves requires courage and ceaseless effort but giardia flagellates "can take it" and reach port safely.

tive tract. The two species of protozoa that live here have evolved interesting methods of meeting this environmental hazard. The giardia flagellate is provided with a sucking disc by means of which it attaches itself to the intestinal wall and the coccidial sporozoon bores its way rapidly into the cells of the intestinal wall where it can grow and reproduce in safety. A giardia flagellate making headway against the peristaltic waves is certainly a winsome thought.

Giardia flagellates are interesting organisms. In the first place they are among the few bilaterally symmetrical protozoa known. Two nuclei are present, four pairs of flagella and two axostyles lying side by side along the center of the body. No oral opening is provided with which to ingest solid particles of food, hence food in liquid form must be absorbed through the surface. When giardias are viewed through a microscope they appear to flutter about with no particular object in view, getting nowhere in particular very slowly. Many of them settle down soon with the sucking disc applied to a solid object. Here they rest except for constant movements of certain flagella which maintain a continuous current about the body, thus insuring an ever-changing environment, a fresh supply of food and the elimination of the products of excretion.

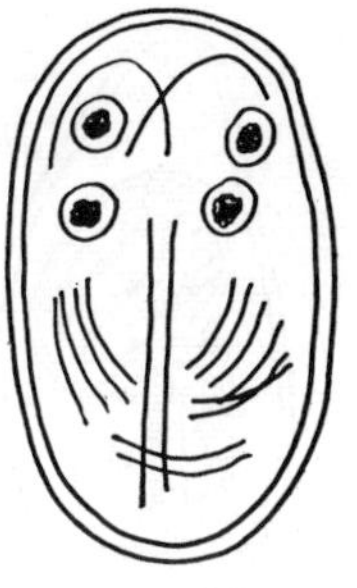

Cyst of the human giardia flagellate

Ordinarily giardias seem to be harmless commensals, but when conditions are such as to favor growth and multiplication, enormous numbers of them are generated; some of these may enter the bile ducts and migrate into the gall bladder. We are not yet quite certain to what degree these flagellates are pathogenic but apparently they bring about cholecystitis and a diarrheic condition.

Egg-shaped cysts are formed by giardias. These are easily recognized by their shape and by the groups of fibers that are visible in both living, and fixed and stained specimens. Division takes place within the cyst and the four, eight or sixteen nuclei that may be present indicate the eventual emergence of two, four or eight young binucleate giardias when the cysts hatch. These cysts pass out of the human body with the excreta and serve to set up infections in susceptible persons who may chance to swallow them with their food or drink.

Giardia infection occurs in about 10 per cent of the general population in temperate regions. Numerous surveys have established beyond doubt that a higher percentage of children are infected than of adults, although no such conspicuous difference in age distribution exists among other human intestinal protozoa. Another peculiarity exhibited by giardias is the fact that in the tropics the incidence of infection is much lower than in northern regions. Since other species of intestinal protozoa are more abundant in the tropics, this unusual condition has stimulated considerable speculation. Biologists can usually suggest some good excuse to account for the eccentric behavior of any organism, but no one that I know of has yet succeeded in giving a plausible explanation of this peculiar circumstance.

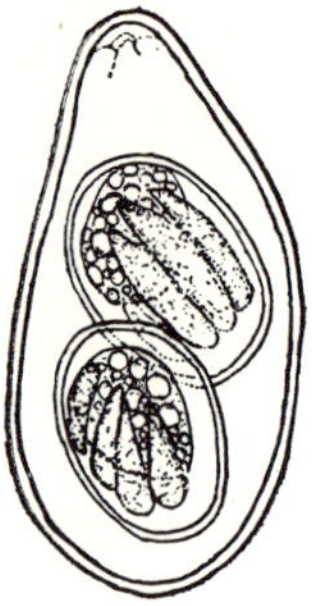

Oöcyst of a coccidium (human) containing two spores each with four sporozoites.

HUMAN COCCIDIOSIS

Human coccidiosis is a rare disease and very little is known about it. For this reason it is of particular interest to parasitologists. We don't know with certainty where the organisms of the disease are located in the human body, and we know nothing about the life cycle of the coccidia while resident in their host. The stage responsible for transmission from one host to another is called the oocyst. This resembles in its significance the cysts of other intestinal protozoa, since the wall of the oocyst is very resistant to the various vicissitudes to which it is subjected when outside of the host's body, and thus is capable of living for a period long enough to enable it to reach the food or drink of human beings in a viable condition. Human beings are so rarely infected, however, and the infection so quickly disappears, with a consequent cessation of oocyst production, that it seems hardly probable that human beings become infected with oocysts from other human beings. The situation seems to require a reservoir host among the lower animals from which man acquires his infection. The identity of this reservoir host is yet to be discovered.

Several types of coccidia have been reported from mammals. Many of them belong to one or the other of two genera, the *Eimeria* or the *Isospora.* Distinguishing members of one genus from those of the other is a simple matter, since the oocysts of *Eimeria* develop four spores within each of which arise two sporozoites; whereas, the oocysts of *Isospora* produce only two spores, each with four sporozoites. The only species of coccidia that parasitizes man belongs to the genus *Isospora* and to the species *hominis.* To be sure a well-known British protozoologist (Dobell) made the ludicrous mistake of describing and naming several new species of coccidia as parasites of man on the basis of oocysts he found in human excrement. Unfortunately for him, he had been guilty previously of criticizing certain other protozoologists with unnecessary severity, hence you can imagine with what glee he was set upon by his colleagues when they were able to establish the fact that these supposedly new species were common parasites of certain fish. The individuals in whose excrement they were found had eaten infected fish and the oocysts had passed completely through the digestive tract without being destroyed. I'm afraid we must admit that none of us is infallible.

The coccidium of man resembles certain species that live in cats and dogs and the best we can do at present is to predict that the human species is located in the same part of the body and passes through the same series of stages in its life cycle, as the other species do in these lower animals. It is a far cry from the simple binary division of *Amoeba proteus* to the complicated series of stages that have had to be developed in coccidia to enable them to live a parasitic existence. However, this life cycle is passed through in a single species of host and cannot be considered very complex when compared with the life cycle of parasites that pass through part of their developmental stages in an intermediate host of another species, for example, the malaria parasite which we shall consider later (see Chapter IX).

LIFE CYCLE OF FELINE COCCIDIA

Briefly stated, when oocysts are swallowed by a susceptible host, the eight sporozoites break out through the wall in the lumen of the small intestine and penetrate the epithelial cells of the intestinal villi. Here they grow at the expense of the protoplasm in the cells of the host and when fully grown divide into 8 or more minute cells called merozoites. The merozoites break out of the cell, attack other cells and repeat the process. Asexual reproduction of this type carried on generation after generation soon increases enormously the number of organisms and hence the number of cells invaded.

After the disease has progressed for a while some of the merozoites become recognizable as of the male type; these microgametocytes do not divide to form merozoites but split up into an immense number of minute motile microgametes comparable to the spermatozoa of higher animals. Other merozoites may develop into a single large body that corresponds to the egg of higher animals. The microgametes escape from the parasitized cell and one of them unites with the macrogamete. This process is similar to fertilization in higher animals. The cell that results from the fusion of a microgamete with a macrogamete develops into an oocyst. The young oocysts break out of the intestinal epithelium into the lumen of the small intestine, are carried by peristalsis into the large intestine and from here out of the body in the excrement of the host.

When oocysts escape from the body they contain a single cell. Soon, however, this cell divides into two and each of these divides into four. Thus are formed eight small cells called sporozoites capable of setting up an infection in any susceptible host that may swallow the oocysts. As is true of other intestinal protozoa, very few ever reach another host, hence immense numbers must be produced to insure the continuation of the race, most of them being wasted by nature in a most profligate manner. Nature in the raw is seen here as a mother who breeds her children by the tens of billions only to see them ruthlessly destroyed. Such inefficiency is unknown among our captains of industry.

PROTOZOA OF THE LARGE INTESTINE

The large intestine supports an extensive and varied population of protozoa. Conditions here are more favorable than in the small intestine. The contents and hence the food supplies, are abundant and well digested; and movement of the contents is much slower than in the small intestine thus making it possible for protozoa to maintain themselves without the aid of special devices for clinging to the surrounding wall. No less than five different species of amebas, four different species of flagellates and one ciliate are known to be normal residents in this habitat. The amebas all seem to be harmless except one species, *Endamoeba histolytica*; this is the villain responsible for amebic dysentery and amebic abscesses of the liver, brain, etc. The flagellates are not pathogenic with the possible exception of two species which, when present in sufficient numbers, may bring about a condition known to physicians as flagellate diarrhea. The ciliate is definitely pathogenic.

The life cycle of most of these protozoa is similar. The actively moving stage, or trophozoite, creeps or swims about in the intestinal contents and ingests solid food particles through the surface of the body or by means of a cell mouth, or absorbs nutriment through the peripheral layer of cyto-

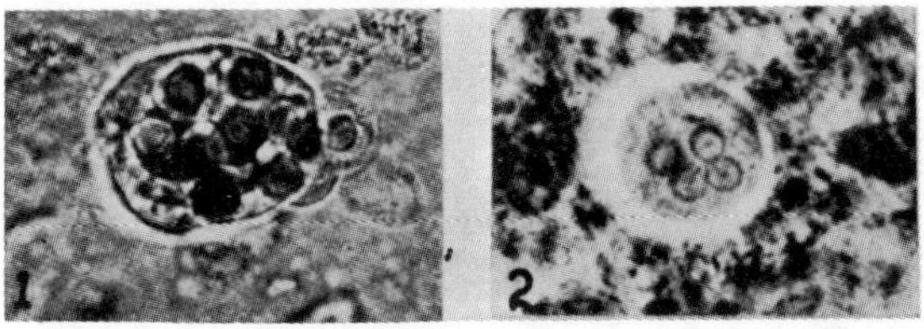

FIG. 63. Dysentery ameba of man. 1. Active specimen filled with red blood corpuscles. 2. Cyst containing four nuclei.

plasm. When fully grown, longitudinal or transverse division takes place according to the species. From time to time, due to stimuli still unknown, certain individuals develop into cysts. These are carried out of the body and are responsible for infecting new hosts: One species, the flagellate *Trichomonas hominis*, has no cyst stage in its life cycle.

AMEBAS OF THE LARGE INTESTINE

The amebas of the human intestine resemble *Amoeba proteus* in their general features. One peculiar difference in all four species is the absence of contractile vacuoles. These conspicuous organelles seem to be very useful in the free-living species, regulating the water content of the body and assisting in respiration and excretion. The parasitic amebas, however, get along perfectly well without them. The human population throughout the world is highly infected with these amebas. About 50 per cent of us harbor the harmless commensal, *Endamoeba coli*; 25 per cent are infected with a second harmless commensal, *Endolimax nana;* 10 per cent with *Endamoeba histolytica,* 10 per cent with *Iodamoeba williamsi* and a very few are inhabited by *Dientamoeba fragilis.* When we consider the large number of other intestinal, blood-in-

FIG. 64. Even the purest of us probably harbors one or more species of protozoan parasites.

habiting and tissue types of protozoa that live in man, it is evident that even the purest of us are probably parasitized by one or more species.

The names of the five species of amebas that live in man are descriptive. *Endamoeba* means amebas within; *coli* refers to the site of infection, the colon, and *histolytica* to the dissolution of tissue by the pathogenic species. *Endolimax nana* moves like a slug, hence *limax* the Latin word for slug, and is very small, therefore the specific name *nana* from the Latin nanus meaning dwarf. *Iodamoeba williamsi* is a species that produces cysts containing a considerable amount of glycogen which stains deeply when treated with iodine; *Dientamoeba fragilis* is a type once supposed to be delicate and fragile, many of which possess two nuclei, hence the prefix *di*.

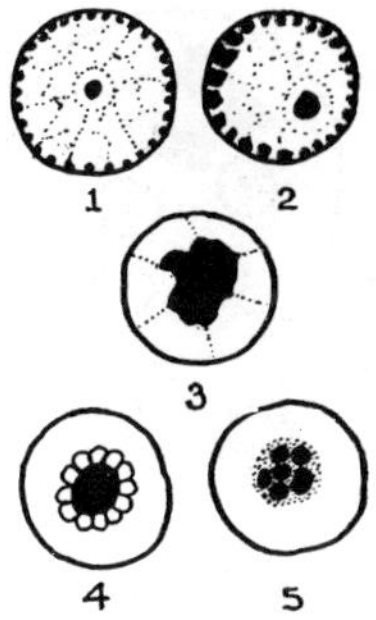

Nuclei of human amebas. 1. *Endamoeba histolytica*. 2. *Endamoeba coli*. 3. *Endolimax nana*. 4. *Iodamoeba williamsi*. 5. *Dientamoeba fragilis*.

Only one of these species is pathogenic, consequently it is necessary to be able to distinguish this one from the others in order to decide whether a patient is suffering from an amebic infection or from some other type of disease. Ordinarily this is done by fixing and staining infected material that has been smeared on a microscopic slide. Size, the structure of the nuclei and the number of nuclei in the cysts are the most important morphological characteristics used in diagnosis. The nuclei of the five intestinal species are presented here in outline. In *Endamoeba histolytica* the nucleus contains a small central particle of chromatin, called the karyosome, and an even layer of delicate chromatin granules around the periphery. In *Endamoeba coli* the karyosome is larger and closer to one side than to the other, and the peripheral granules are coarser. The nucleus of *Endolimax nana* has no peripheral chromatin granules the chromatin being massed into one or several irregular blobs that are attached to the nuclear membrane by delicate strands. *Iodamoeba* possesses a nucleus containing one large mass of chromatin surrounded by a layer of colorless granules. And the nuclei of *Dientamoeba fragilis* possess half a dozen rather heavy chromatin granules held together in a colorless matrix. These characteristics are typical and constant, but one must have considerable practice before he can employ them with confidence. Good eyes, a steady hand and an abundant supply of scientific patience are required for work in this field.

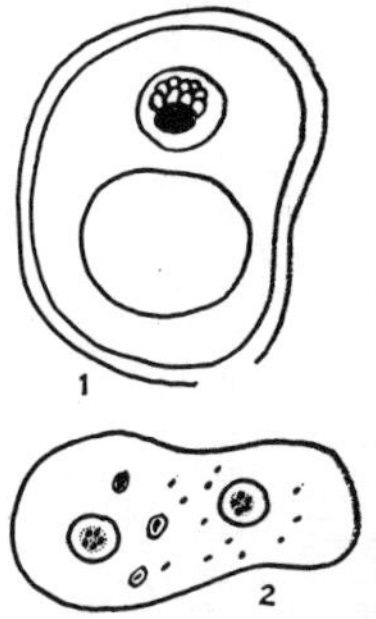

Human amebas. 1. Iodamoeba, cyst. 2. Dientamoeba, with two nuclei.

The life cycles of these intestinal amebas are simple. Reproduction is entirely asexual. The active amebas, the trophozoites, divide in the colon by binary fission just as *Amoeba proteus* does in fresh water. In all except *Dientamoeba fragilis*,

Cyst of *Endamoeba histolytica* of man

cysts make their appearance from time to time. When encystment occurs the ameba extrudes all food bodies and becomes more or less spherical. Then it secretes a cyst wall about itself which is resistant to various injurious agents in the environment outside of the body. The nucleus in the cysts of *Iodamoeba* do not divide, but in the cysts of *Endamoeba histolytica* two nuclear divisions occur, hence the mature cysts contain four nuclei. Cysts of *Endolimax* likewise possess four nuclei, but in *Endamoeba coli* a third nuclear division takes place and mature cysts are therefore provided with eight nuclei. In *Iodamoeba,* cyst formation appears to be solely for purposes of protection, whereas in *Endamoeba coli, Endamoeba histolytica,* and *Endolimax nana,* asexual reproduction occurs within the cysts. If cysts are formed by *Dientamoeba fragilis* they must be exceedingly rare, since many observers have looked for them but none have been recognized with certainty.

FLAGELLATES OF THE LARGE INTESTINE

Of the four species of flagellates that live in the large intestine of man two are relatively widespread and the other two seem to be rare. The latter are of interest only to professional protozoologists, hence we may dismiss them with nothing more than our blessing. The other two have been found in human beings in many parts of the world in about ten per cent of those examined. Both of them have been accused of causing intestinal disturbances but this has not been definitely established. The intestinal trichomonad, *Trichomonas hominis,* resembles in structure the species that occurs in the mouth and only a trained observer can distinguish between the two. Fortunately the two species live in different parts of the body and hence can be identified by their habitat. The other intestinal flagellate goes by the name of *Chilomastix mesnili.* This species is a trifle larger than *Trichomonas hominis,* lacks the backbone-like axostyle, and the undulating membrane, but is provided with a cyst stage in its life cycle. This cyst is easily identified because of its lemon shape, there being no other species with a cyst of this type. The relations of these two species to their human hosts will be discussed later.

Cyst of chilomastix flagellate of man

OLD MAN BALANTIDIUM

The largest of the protozoa that live in the large intestine of man is the ciliate, *Balantidium coli*. Balantidium looks very much like the free-living ciliate, paramecium, and does not

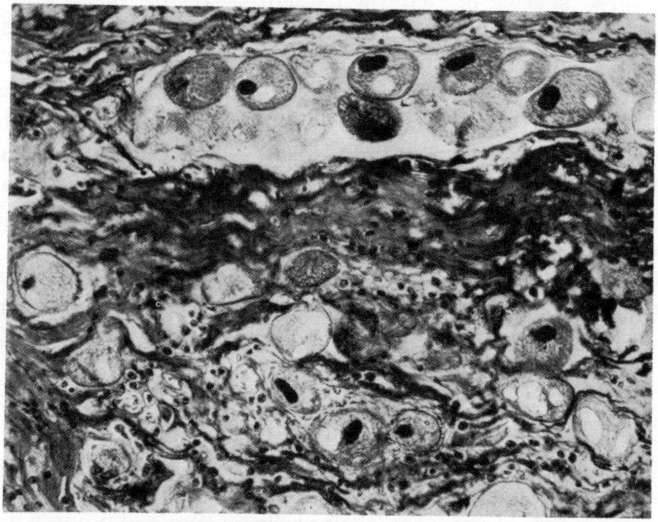

FIG. 65. The balantidial ciliates in this section of the wall of a human intestine can be identified by the large black nucleus in each.

seem to have become modified to any considerable extent by its parasitic habits. It has a cell mouth with which it ingests solid particles of food, a cytopyge near the posterior end through which undigested particles are discharged from the body, a large macronucleus and a smaller micronucleus, and two contractile vacuoles. In certain hosts cysts are formed, but in others the cyst stage in the life cycle appears to be of rare occurrence. Most of the human beings that are infected with this species are in the carrier condition and do not exhibit symptoms, but occasionally a patient may suffer from balantidial dysentery which is very similar in its character to amebic dysentery.

BLOOD-INHABITING PROTOZOA

The blood-inhabiting protozoa that are known to live in man are of three main types and comprise three species of trypanosome flagellates, three species of leishmania flagellates and three species of malaria parasites. So far as is known these protozoa do not pass directly from man to man but depend for their transmission on an intermediate host. We therefore share these infections with tsetse flies, triatoma bugs, sand flies, and mosquitoes. Obviously diseases due to this group of parasites cannot exist except where the intermediate hosts live. On this account blood-inhabiting protozoa are restricted in their distribution, whereas intestinal protozoa occur in every inhabited country in the world.

The blood-inhabiting protozoa of man will be discussed fully in succeeding chapters and their consideration is, therefore, reserved until then.

TISSUE PROTOZOA

One tissue parasite has attempted to "muscle in" where it "can't take it" since it has never succeeded in adapting itself satisfactorily in man. This is *Sarcocystis*, a type that forms large cysts full of spores in the muscles, hence the name which is derived from Greek words meaning flesh and bladder or cyst. Much is still to be learned regarding this peculiar denizen of the protozoan inner world (see page 70). Only about a dozen cases of sarcosporidiosis have been reported from man, the parasites usually being found at the time of autopsy. Unlike the infection in sheep and mice, the organisms do not spread throughout the body and do not cause serious injury. How human beings become infected is unknown. Certainly there are not enough diseased persons to transmit the organisms from one human host to another. The suggestion has been made that the human sarcosporidia are normal parasites of certain lower animals which have been swallowed by man and have succeeded in maintaining themselves in the human body. There seems to be no way for them to escape, hence they are imprisoned in a "blind alley" and are "all dressed up but with nowhere to go."

TYPES OF MONKEYS

According to Pope "The proper study of mankind is man." If we accept this dictum, the protozoologist may well be proud, because he has learned more about the protozoa of

FIG. 66. Miss Sarcocystis lives in a "blind alley" among the muscle fibers and is "all dressed up but with nowhere to go."

man than he has about those of any other animal. Our knowledge of the protozoa of monkeys, however, has been very meager until quite recently. The similarity in structure of monkeys and men should have attracted the attention of parasitologists much earlier, as it did the anatomists, but now, at any rate, their possibilities seem fully appreciated and information about monkey protozoa is piling up with great rapidity.

Everybody knows a monkey when he sees one and thoughtful people are glad that they were not born at the monkey stage of evolution. Monkeys, on the other hand, seem to derive a great deal of enjoyment from life, and no doubt congratulate themselves that they have not developed into men. It would add to their satisfaction if they could realize how many scientists spend their lives investigating them.

At one time in the history of zoology a scientist covered in his studies the entire animal kingdom, but zoology has become more complex as our knowledge of animals has grown and it has become necessary for specialists, who continually learn more and more about less and less, to devote their entire time to the study of monkeyology. Monkeys are also studied by anthropologists; anthropology has been defined as the science of man embracing woman. These two groups of scientists, the zoologists and the anthropologists, have divided the more than one-hundred species of monkeys into eight families. Those who are not specialists delude themselves into thinking that a monkey is a monkey, but zoologists recognize monkeys that are aye-ayes, tarsiers, lemurs, marmosets, capuchins, titis, baboons, macaques, gibbons, orang-utans, chimpanzees, and gorillas. There are some who would even include certain individuals of the genus *Homo* in this group.

PROTOZOA OF CAPTIVE AND WILD MONKEYS

For many years most of the monkey protozoa examined came from animals that had been kept in captivity, where they might have become contaminated with human protozoa. But many wild monkeys have been studied during the past few years and have yielded just as many parasites as their imprisoned relatives. Where these wild monkeys got their protozoa is an intriguing problem. Infections at the present time are no doubt transmitted from one parent to another and from parents to offspring much as human protozoa are distributed among the members of a family. This method of transmission can likewise be imagined to have occurred generation after generation up from the dim and misty corridors of time. If we believe in organic evolution, the probability seems unquestionable that the protozoa of a living species of monkey must have been thrust upon it by the species out of which it evolved. If we accept the principle of special creation, then men as well as monkeys were from the very beginning endowed with a complete set of parasites, the Garden of Eden was populated by other living things besides Adam and Eve, apple trees and serpents, and, so far as parasitic protozoa are concerned, nothing could be more certain than that "Adam Had 'Em."

FIG. 67. Whether you believe in organic evolution or in special creation no doubt exists regarding the presence of protozoa in the Garden of Eden because Adam had 'em.

COMPARISON BETWEEN HUMAN AND MONKEY PROTOZOA

Protozoologists have damaged their eyesight dreadfully in their attempts to see differences between the protozoa of monkeys and men. Some of the protozoa appear to be more human than others, just as some of our friends seem more like monkeys than they do like men. Many human protozoa have been inoculated into monkeys and have apparently been able to grow and multiply there successfully. This does not prove that man is infected with monkey protozoa, but indicates, at least, that the human species must be very closely related indeed to the monkey species.

At the present time we do not know of any structural characteristics that can be used to distinguish between most of the species of human and monkey protozoa. This applies to the ameba and flagellate of the mouth, the giardia flagellate of the small intestine, the five species of amebas, three of the species of flagellates and the ciliate that live in the large intestine, the trichomonad flagellate that inhabits the genital tract, and two of the trypanosome flagellates and the three malaria parasites that live in the blood.

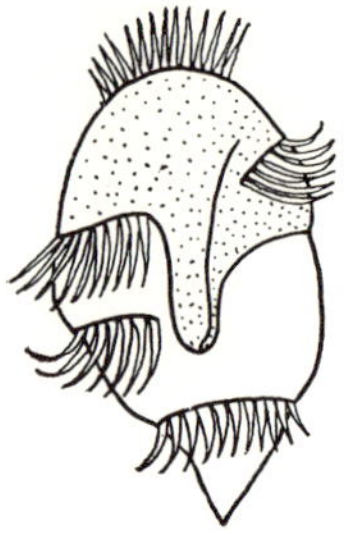

Troglodytella, an intestinal parasite of chimpanzees and gorillas.

This means that eighteen of the twenty-five species of protozoa that are known to live in man are indistinguishable from eighteen of those that have been reported from monkeys. The other seven human species may live in monkeys somewhere but have not yet been discovered. Besides the eighteen species mentioned, monkeys are known to be parasitized by a large ciliate, *Troglodytella,* by two other species of ciliates and by several other species of protozoa whose specific identity is still in doubt. Thus about seven monkey species do not live in man and about the same number of human species do not live in monkeys.

Whether the protozoa of monkeys and man that cannot be distinguished from one another really belong to the same species is difficult to decide. Recent studies of the balantidial ciliates of pig, monkey, and chimpanzee suggest that differences may exist that are very obscure. Balantidia from time to time conjugate two by two much as paramecia do. After the conjugants separate, a complicated reorganization of the nuclei takes place in each exconjugant. A study of this process of nuclear reorganization has revealed differences of a fundamental nature between the balantidia of the pig and chimp, on the one hand, and of the guinea-pig on the other. The only conclusion possible is that the guinea-pig is inhabited by balantidia that are specifically different from those in the other two animals, although they appear to be alike superficially. We don't know whether the human balantidia favor those in the guinea-pig or those in the chimp and pig, but we are betting on the latter. Other protozoa of monkeys and men, when very carefully studied may disclose similarly obscure differences.

MALARIA PARASITES OF MONKEYS

The frolicsome antics of monkeys ought to discourage the bites of mosquitoes, which are obliged to settle before they can successfully operate their stingarees, but in nature our sprightly relatives suffer severely from malaria and must therefore have served as meal tickets for those murderous females that fly on two wings and sing, what is often a death knell, before they sting. Monkeys live in tropical and semitropical regions where mosquitoes abound during the entire year and apparently all species of monkeys throughout their geographical range are subject to malaria. As early as 1898 malaria parasites were discovered by Koch in African monkeys. Of special interest are the three species that are known to occur in chimps and gorillas and that resemble the three human species. All of these have been observed in a single chimpanzee. Various attempts have been made to infect men with these organisms and to infect chimpanzees and other species of monkeys with the human parasites, but, although certain positive results have been reported, the probability is that none of the infections were actually due to the parasites injected into the several hosts. The conclusion is that the species in chimps and men are probably different in spite of their morphological similarity.

The lower monkeys, however, seem to be inhabited by malaria parasites visibly different from those in men. For example, *Plasmodium kochi* in African monkeys, *P. knowlesi* in a species from Malaya, *P. cynomolgi* from another species from Malaya, *P. semnopitheci* in monkeys from Assam, *P. inui* in a species from southeastern Asia and the Dutch East Indies, and *P. brasilianum* in Cebus monkeys from South and Central America. Recently infections in human beings with *P. knowlesi* have been reported, and this species of monkey malaria is now being used in the treatment of general paralysis in man. (See Chapter X.)

COMPARISON OF HUMAN AND ANIMAL PROTOZOA

It we compare the protozoa of man with those that live in any wild or domesticated animal other than monkeys, we behold an entirely different picture. In most instances we may encounter protozoa belonging to the same genera but to obviously different species. Thus in the rat we note an intestinal ameba (*Endamoeba muris*), a giardia flagellate (*Giardia muris*), several intestinal trichomonads (*Trichomonas muris*, *T. parva* and *T. minuta*) and a trypanosome (*T. lewisi*) all visibly different from species of the same genera that inhabit men, as well as species belonging to genera that are not represented among human protozoa, such as three species of coccidia (*Eimeria miyairii*, *E. separata*, and *E. carinii*), and a blood-inhabiting sporozoon (*Hepatozoon muris*). Similar comparisons could be made with other species of animals, not only with mammals but also with birds, reptiles, amphibians, fish and even inverte-

FIG. 68. Madame Monkey displays her rosary of parasites to Mrs. Rat and boasts because they resemble those of Mrs. Homo.

brates. Monkeys, therefore, may with good reason boast about the character of their parasitic protozoa.

The biological significance of this situation seems evident when we consider the fact that animals closely related, according to their arrangement in the animal series on the basis of organic evolution, are known to be inhabited by similar genera and species of parasites. Or, stated in another way, the greater the similarity between the parasites the more closely related are the hosts in which these parasites live. The extraordinary resemblance between the protozoa of monkeys and men can lead to but one conclusion, and that is, that men are more closely related to monkeys than to any other type of lower animal. Parasitism may therefore be added to such subjects as classification, comparative anatomy, embryology, geographical distribution and paleontology in support of the theory of organic evolution.

We are forced to conclude, therefore, that men and monkeys are brothers under the skin, since both support a similar lot of microscopic descendants of the same original stock. Fortunately monkeys are unaware of the billions of jazz babies gnawing at their vitals or they wouldn't be so cheerful.

FIG. 69. Monkeys are full of parasites but fortunately don't know it or they wouldn't be so cheerful.

The poem with which we conclude this chapter carries a title which is itself a poem—probably the shortest poem in the English language, or any other language, including the Scandinavian—the title is "Adam Had 'Em."

Amebas abound in your kisses
And flagellates lurk on your lips,
Your bowels are all swarming with microbes,
B. coli, Giardia and sich.

Last night as you lay on your pillow,
A rhumba was danced in your blood,
Corpuscles were broken asunder,
And trypanosomas made love.

You'r nought but a mass of corruption
Passed down from a simian tree
To Adam and Eve and their offspring.
Who says we are equal and free?

Chapter VI

MONOS DE LA JUNGLA

MUCH of the labor required of protozoologists is of the "busy work" variety, namely, hunting for parasites, preparing microscopic slides for examination, caring for laboratory animals, etc., but a protozoologist who wishes to travel and visit strange places can always find a legitimate excuse to do so. Among the writer's most interesting experiences have been excursions into the jungles in the Philippines and in Panama for the purpose of obtaining protozoan parasites from wild monkeys. One does not need to travel very far from Manila to encounter jungles that have never been explored by white men and are well populated with monkeys—and pigmies. The monkeys are, of course, harmless to man, but the pigmies still use blow guns and poisoned arrows and at times may become a real menace.

MONKEY HUNTING IN THE PHILIPPINES

Monkey hunting in the Philippines cannot be considered very exciting or much of a sport, especially when the principal object is to secure monkeys, and shot guns are used for this purpose. But to one who has spent most of his life in comparatively highly civilized regions in the temperate zone, a day in a tropical jungle always provides a definite thrill,—the hot sun and humid air, the lush vegetation, the springy floor of the forest, the drip, drip of water from the leaves after the daily rains, the "talking" of parrots as they fly by overhead, and the many strange voices of the jungle inhabitants furnish a background quite foreign to that to which one is accustomed.

The character of one's companions plays no small part in the pleasure of a jungle excursion. Our hunting party in the

Philippines included a Chinese student from Soochow, a Filipino assistant, and several Filipino hunters engaged on each expedition from among the natives near whose homes the area to be explored was located. There was thus plenty of variety among the personnel.

We usually set out from Manila by automobile at about six in the morning when the air was comparatively fresh and cool. Virgin jungle was ordinarily reached in three or four hours.

FIG. 70. A group of monkey hunters in the Philippines. The author is holding up a monkey he has just shot. At the left is his Chinese assistant, Mr. Chu. The other three are Filipino hunters.

Then my Filipino assistant, Manipol, would round up some native hunters, who always seemed delighted at the chance of a day's sport, and would secure a guide and the hunt would begin. Some jungle is so dense that penetration is practically impossible but other regions are surprisingly free from underbrush. The wise protozoologist selects the latter as the scene of operations.

We practiced two methods of collecting monkeys; one was

to find the monkeys and the other to allow the monkeys to find us. One can travel through a clean stretch of jungle without making much noise, but a little noise is enough to attract the attention of monkeys and to send them silently into some place of concealment. But the natives know their habits and could see them among the leaves in the tops of the trees. Usually our shots were of the real shot-gun variety, that is, we blazed away at the spot where the monkey ought to be, trusting to luck that he was there and would receive a mortal wound.

Often our shots were wasted in a dense clump of leaves that concealed nothing at all, but the fact that we were able to report the examination of forty-four wild monkeys indicates that we were sometimes successful. Occasionally the hidden monkey became too nervous to remain quiet and would start off through the tree tops on a wild scramble to get out of range. This always meant going "over the top" for us, since we had to set forth on an equally wild scramble over the floor of the jungle in order to catch up with the fleeing monkeys. We lost about as many as we bagged on these occasions, became outrageously begrimed as a result of frequent headlong dives into the murky soil when our feet encountered an exposed root, and usually collected a supply of unwelcome thorns with which jungle vegetation is so abundantly provided, as well as other bruises of various sorts.

Waiting for the monkeys to find the hunter may not yield as large a bag but is not as exhausting and furthermore enables one to contemplate the jungle at his leisure. The trees of the jungle pass through stages of youth, maturity and old age in rapid succession and the rotting trunks of fallen patriarchs were everywhere available to sit on. The most characteristic sound in the jungle, at least during the rainy season, is the constant dripping of water. Birds and larger animals are surprisingly scarce. Where open spaces occur and a bit of sunlight is allowed to penetrate, one may observe gorgeous butterflies fluttering aimlessly about. The base of the jungle is by day too light for night-hunting animals and too dark for diurnal inhabitants and hence is practically free from observable animal life. Daylight activities are mostly confined to the tops of the trees, which are usually so far above that little can be accomplished even with powerful field glasses. What an inspiring experience it would be for a naturalist if he could wander about among the jungle tree tops with the ease and agility of a monkey. From the naturalist's standpoint it may seem like a mistake to have evolved so far away from our simian ancestors as to loose a prehensile tail, indifference to high places, and arboreal habits.

The thrills of monkey hunting do not end with the capture of the monkey, at least for protozoologists, since the question as

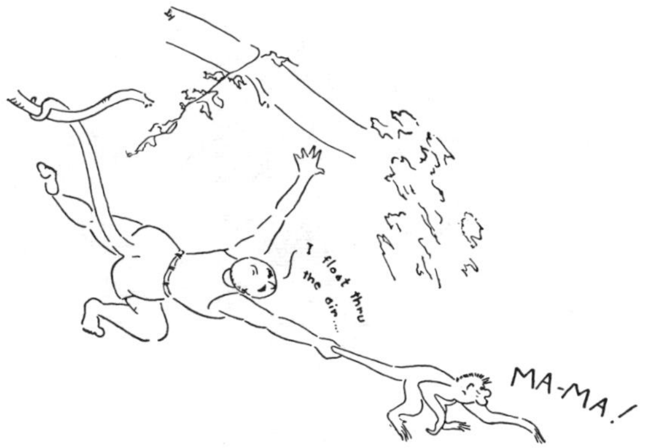

FIG. 71. We could "bring 'em back alive" if we had a prehensile tail and could wander through the tree tops.

to what protozoa we might find in our prey was never out of our minds. So home we would travel in our automobile as fast as possible, which was not very fast, due to slow-moving carabao carts that often blocked the road and to the habit of the natives of loitering all over the right of way. Arrival at the laboratory was followed by a hasty unloading of the monkeys, a shot of coffee and bolt of sandwich, and then the long-anticipated autopsies.

HUNTING FOR PROTOZOA IN MONKEYS

Your happiness probably does not depend on knowing how a protozoologist autopsies a monkey but this is an adventure that almost anyone should find of interest. And, by the way, none of us ever felt that we were killing anything but a wild animal when we shot a monkey; there is no similarity between a jungle monkey and a human baby such as some writers would have you believe. Furthermore, dissecting a monkey seemed in no way different from cutting up a dog or cat.

As noted in the preceding chapter, the protozoa of monkeys resemble closely those of man not only in appearance but also

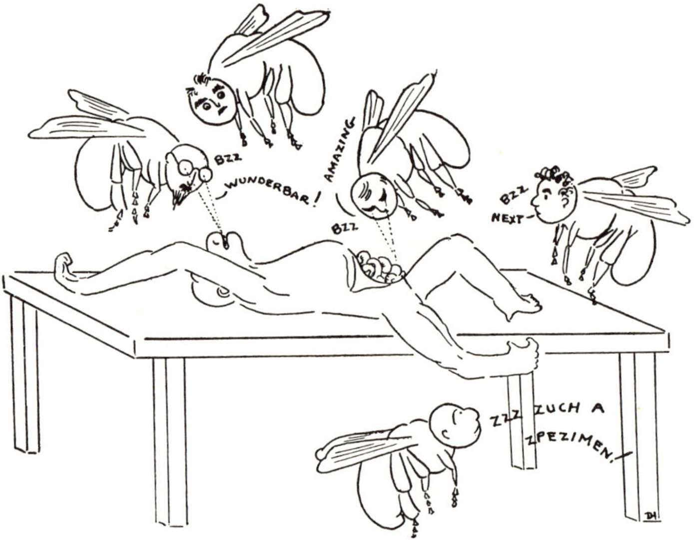

Fig. 72. When protozoologists have a monkey to autopsy they are as busy and excited as little bees.

as regards their localization within the body. We first scraped away material from around the bases of the teeth placed it on a microscopic slide, diluted it with saline solution, placed a cover-glass on top, and then examined it under both low and high magnification. Minute protozoa, if moving, can be seen with low powers of the microscope and, since a much larger field is visible with low power, the preparation was scrutinized first for moving organisms. Then the higher powers were employed with the hope of finding protozoa that might be quiescent. Thirty-seven of the forty-four wild monkeys examined were found to be infected with the oral type of ameba and thirty-six with the oral trichomonad flagellate. Those who have worked extensively with monkeys in captivity tell us that monkeys do not kiss each other, hence the transmission of these protozoa of the mouth must take place in some other way. Monkeys are notoriously dirty animals and probably oral protozoa are distributed among them as they slobber on their food in their struggles for the choicest morsels. Our discovery of oral trichomonads in monkeys was the first record of this species in this type of host.

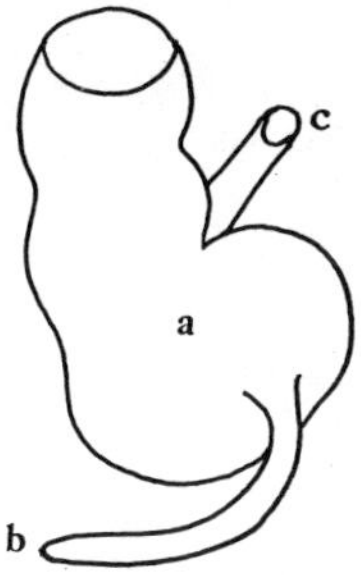

Cecum and appendix of man. a. Cecum. b. Appendix. c. Small intestine.

We next opened the abdominal cavity and removed the entire intestine. Samples were taken and examined from various regions of the small intestine and the intestinal wall was carefully studied to see if any injuries that might be due to protozoa were visible. Giardia flagellates were observed in six monkeys, but no coccidia, which are rare inhabitants of the small intestine of man.

The cecum of many animals is a blind pouch into which the small intestine opens. In certain species a wormlike, tubular sac projects out from the cecum; this is the vermiform appendix, which in man is from three to six inches long and about three-tenths of an inch in diameter. The cecum is a favorite residence of intestinal protozoa some of which may enter the appendix. The movement of the cecal contents is comparatively slow, hence protozoa maintain themselves there with the least possible expenditure of energy. An abundant food supply is also available here at all times. Seven species of protozoa were found in this part of the monkey digestive tract. The friendly relations of these species is indicated by the fact that three of them were living together happily in twelve of the monkeys, four species in ten monkeys and five in three of them. We do not claim to have detected all of the species present hence these are minimum numbers. As a matter of fact, we became more expert in finding certain species as the work proceeded; for example, we discovered that balantidial ciliates were often present in one particular pocket-like swelling where movement of the contents was probably less rapid than in other regions of the cecum.

In the cecum of these monkeys the peculiar species of ameba known as *Dientamoeba fragilis* was recorded from monkeys for the first time; in fact, this type of ameba has been reported from only one other host, namely, man. Intestinal flagellates

FIG. 73. Protozoa live in the cecum contentedly because one never trumps another's ace.

were quite common as well as balantidial ciliates. Dysentery amebas (*Endamoeba histolytica*) were noted in ten of the forty-four monkeys, but no injuries resulting from their presence were discovered.

Only sporadic efforts were made to obtain blood-inhabiting protozoa from these monkeys, since, by the time we had completed our search for intestinal species, we were all completely exhausted and quite ready for a hot bath and a good night's sleep. We did examine vaginal smears from the sixteen female monkeys in our collection and detected trichomonads in two of them.

MONKEY HUNTING IN TROPICAL AMERICA

One of the best hunting grounds for monkeys in tropical America is the so-called Coto region that is located where Panama and Costa Rica join. Here a party of biologists from Johns Hopkins, Harvard and Yale together with the director of the Gorgas Memorial Laboratory in Panama City spent a few weeks during the summer of 1932 collecting and studying wild monkeys under the auspices of the Gorgas Memorial Laboratory and the United Fruit Co. The Fruit Co. put up a camp for us on the La Vaca river several miles beyond their most distant banana plantation and transported us and our baggage to this camp via narrow-gauge railroad and muleback.

Life in this camp was distinctly pleasant. We brought two hunters with us, one a Yaqui Indian and the other a mixture of Indian, Spanish and Negro. With much effort a negress who could supply our simple needs as regards food, also made the trip. Shelter consisted of a large hut with a rain-proof thatched roof but no sides. One end of this served as a store house and dining room and the other for sleeping quarters. Each of us had a bed consisting of a strip of canvas stretched between posts three feet from the ground. Each bed was covered by a canopy of cloth netting to protect us from malarial mosquitoes. Every night we got under our canopies immediately after an early supper, that is, at about six o'clock, and remained there until after daylight the next morning. Since malarial mosquitoes become active at twilight and remain active until after daybreak, we escaped not only their stings but also infection with the malarial organisms that they carry.

Our daily routine was somewhat as follows. The two hunters started out at daybreak after monkeys. Breakfast was served at about seven o'clock. Between 7:30 and 10 we were engaged in various activities according to the character of the investigations we were interested in. The writer and his assistant worked up the records of the previous day and stained slides. Occasionally we set forth into the jungle to see if we could locate any fresh meat, such as an iguana lizard or a

FIG. 74. The author arrives in camp with two adult red spider monkeys that he has just shot in the steaming jungle, and with a living baby red spider. (Rio La Vaca Camp, Panama.)

jungle turkey (curassow). And once in a while we brought back a monkey or two. The behaviorist usually entered the jungle immediately after breakfast loaded with camera and field glasses and devoted the day to a study of the activities of monkeys in their native haunts. The anthropologist was making a collection of skulls and spent practically all of his time cleaning bones. The anatomist made preliminary studies of the various organs he was preserving for future investigation, and the director had his hands full taking care of the rest of us.

At about 10 o'clock the hunters returned with from three or four to six or eight monkeys and from then on our camp was as busy as a bee hive. The anthropologist numbered, weighed

FIG. 75. The candid camera finds its way into the jungle where a red spider monkey is yielding up its parasites all for the sake of science (Rio La Vaca Camp, Panama).

and measured each monkey. The director was particularly interested in the food of monkeys and therefore made an examination of the contents of the stomach and intestines; he also prepared slides of the blood of each animal and collected parasitic worms from the intestine and body cavity. Then the intestine was turned over to the protozoologists who examined the walls for injuries that might be due to protozoa and the contents for the organisms themselves. The anatomist then proceeded to dissect out certain internal organs in which he was interested. Finally the carcass was delivered to the anthropologist who made further measurements and then began the drudgery of cleaning the skeleton. Very little of the monkey was left when we had all completed our tasks; we even surpassed the meat packers, who are said to use all of the pig but his squeal, since our behaviorist noted the character of the monkey's activities before he was killed and has since published a monograph on the subject.

FIG. 76. The protozoologists of the expedition are finding a new ciliate which they name after two of their colleagues, *Taliaferria clarki* (Rio La Vaca Camp, Panama).

The protozoological results of our studies are similar to those obtained in the Philippines. Many of the one hundred monkeys examined were found to be infected with one or more intestinal protozoa. With two exceptions these could not be distinguished morphologically from similar species that live in man. A new genus and species of ciliate which was named *Taliaferria clarki* and a new species of *Balantidium*, named *B. wenrichi*, were discovered in the red spider monkey. All together seventy-two red spider monkeys, eight white-faced (organ grinder) monkeys, and twenty three titi monkeys were collected and studied.

One of the most novel features of our camp was the complete isolation from all man-made noises,—no automobiles, no steam engines, no phonographs, no radio, no whistles, no hard-heeled shoes pounding on cement sidewalks, and no talking laughing and quarreling except what we did ourselves. At night the gurgling of the water in the river was constantly in our ears, interrupted now and then by the scream of a jungle animal,

FIG. 77. White-faced monkey

perhaps directed to a mate, or a rival, or the death cry of a hunting animal that had been captured by a larger foe. Night-hunting with jacklights was indulged in to a certain extent. On one hunt we shot a tiger cat which on examination proved to be badly damaged by tuberculosis, and during another expedition one of us was just about to make a lunge for what he supposed was a large toad whose eyes were shining when he was pushed aside by the guide who shot the creature which proved to be an enormous fer-de-lance, one of the most deadly of all snakes.

CHIMPANZEES IN FLORIDA

The inhabitants of Florida might or might not be pleased to have any part of their state called a jungle, but at any rate they delight in tropical vegetation and enjoy a higher temperature than northerners are accustomed to. This will perhaps be accepted as an excuse for including in this chapter an account of the protozoological study of the captive chimpanzees maintained at the Anthropoid Station at Orange Park, near Jacksonville, Florida. Florida was selected as the site of this Station because mild weather appears to be more favorable for the health and reproductive activity of chimps than snow and ice.

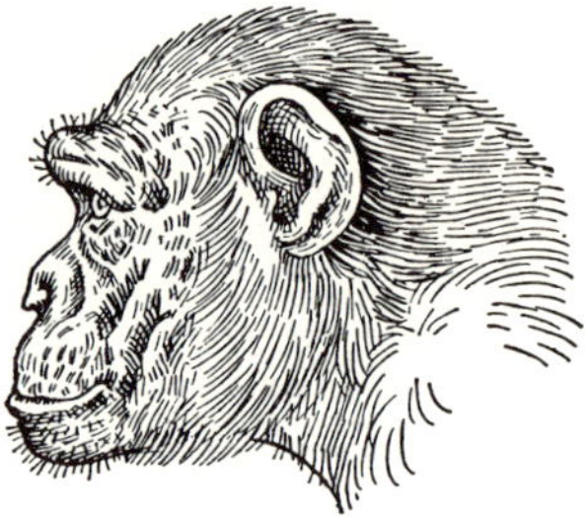

FIG. 78. Head of chimpanzee

Setting up a laboratory in the midst of the peaceful Florida landscape and populating it with noisy chimpanzees whose tremendous bellowing can be heard for a mile or more was certain to interest the natives in that neighborhood. Why anyone should go to so much trouble and expense for the sake of a few chimpanzees was a constant source of wonder. The attitude of the Florida "crackers" was indicated by a conversation that was overheard between two of these residents. One said to the other, "What are they trying to do with all those apes." The other replied, "I reckon they're trying to train 'em." "What's the good of training 'em?" said the first cracker, "As soon as they're trained, they'll free 'em and give 'em the vote."

At this Anthropoid Station were kept a large group of adult chimpanzees and a number of young of various ages some of whom had been born on the premises. No one can approach an adult chimp with safety, therefore strong cages were built in which they lived. Each cage was provided with an enclosed annex into which the chimps would go from time to time

FIG. 79. Brother Chimp shows his appreciation of the New Deal

for a bit of rest and quiet. When the door into the annex was closed and fastened the cage could be cleaned out. Through the kindness of the director of the station and his helpers we were able to obtain ten intestinal samples from each of twenty-two chimpanzees. Parts of these samples were inoculated into culture media and the rest examined as fresh preparations.

This represents the most extensive survey of the intestinal protozoa of chimpanzees to date.

One of the most interesting of the protozoa found in chimps is the large ciliate known as *Troglodytella abrassarti*. This species was first described from a chimpanzee in 1912, and later from a gorilla. It was present in six of fifteen animals examined in Baltimore but was absent from the twenty-two in Florida, although several of these were known to have been infected with this species before being taken to Florida. No troglodytellas have ever been reported from man.

Of the seven species of intestinal protozoa found in the Florida chimps, four were recorded from these apes for the first time; these were giardia flagellates from the small intestine, retortamonad and hexamita flagellates from the large intestine and trichomonad flagellates from the vagina. Balantidial ciliates and two species of amebas were also present. All of these protozoa, except hexamita, occur in man and those we examined from the chimps were morphologically indistinguishable from the human species. Thus once more are we reminded of our close kinship to the higher apes.

Field work of the type described in this chapter is not only a delightful adventure when in progress but is also pleasant in retrospect. As we sit before the fireplace on winter evenings after a day devoted to more or less ineffective attempts to guide students along the narrow path that leads to a higher degree in protozoology, our thoughts are apt to stray lovingly to those good old jungle days and to take poetic form somewhat as follows.

In Filipino's tropic clime
Where everything entrances,
The monkey and the tarsier
Free wheel among the branches.

In Panamanian wilds as well
The spider and the owl
Are monkeys full of parasites,
Ye Gods, just hear them howl.

Monos de la Jungla

The titi too luxuriates,
Capuchins and their ilk
Make merry in the jungle,
Except those that we kilt.

And now that rocking chair's got me,
And old age obscures my gaze,
The dripping jungle with its apes,
Good bye, those good old days.

Chapter VII

TELL ME WHAT YOU EAT

The time has come, the author thinks,
To talk of what you eat:
Of steak—and starch—and anaerobes—
Of everything that's sweet—
Why vegetarians suffer so,
Because they won't eat meat.

AS A rule we pay very little attention to our digestive system unless we have a tummy ache, but even without this affliction certain of the more introspective zoologists have interested themselves in the problems presented by our intestinal fauna. The species of protozoa that may live in our digestive tract and blood stream have already been described and some phases of their relations to the human body discussed. We have also learned that certain types of animals, such as the termites, cannot exist unless they are parasitized with intestinal flagellates, and that a very large percentage of other animals may be inhabited by intestinal protozoa.

Ordinarily we think of the normal condition of the human body as that free from parasites, yet in many localities 100 per cent of the population are infected with certain species and over 50 per cent of these may be infected with other species. In such localities it might be correct to conclude that those who are parasitized are normal and anyone who is not infected abnormal, or at least, peculiar.

Nowadays it is customary for criminals to tell their life stories to the world before their execution. Lack of a voice prevents the protozoa that are guilty of invading the human body from broadcasting the lurid details of their wickedness;

for this reason I am going to assume the rôle of a ghost writer willing and ready to reveal what the scientific fraternity has been able to uncover about them. But first you are asked to imagine the human body as a world consisting of a league of protozoan nations, all friendly toward one another but constantly battling with their peculiar environment. Human protozoa are carried from place to place within the body by

FIG. 80. An introspective zoologist becomes interested in his intestinal protozoa.

way of the digestive tract, glandular ducts and the circulatory system much as we travel in railroad trains, automobiles and airplanes. Just as nations on earth are restricted to more or less definite areas, so human protozoa are localized in limited regions of the body. Just as we spend our time maintaining ourselves and raising offspring to prevent the human race from dying out, so protozoa are constantly engaged in obtaining nourishment, warding off enemies and reproducing.

Reaching a human body in which to live is a much more difficult feat for protozoan parasites than house hunting is for us, since protozoa have adopted a passive attitude and depend entirely on their host's carelessness for coöperation. Two main entrances invite the visits of these microörganisms. Intestinal protozoa can invade the human body only through the mouth. Blood-inhabiting protozoa must reach the circulatory system; into this they are inoculated through the piercing beak of a blood-sucking insect. Since blood capillaries penetrate to all parts of the body, there is no area of human tissue that is not visited by blood-inhabiting protozoa. However, different species of human protozoa may live in different tissues or organs and some live a more aquatic existence in the blood plasma than others that are located within the cells. Thus a variety of habitats are available, each with its own peculiar characteristics and each with its burden of parasites to entertain. All this tends to minimize the rivalry among the different species, and, as already noted, results in the absence of antagonism.

PROTOZOA OF THE HUMAN MOUTH

We may devote our attention first to the two species of protozoa that live in the mouth. Some of the trials and tribulations of these badly harassed protozoa were described in Chapter V. Life for them begins after the host finishes his dinner at eight in the evening and ends when he brushes his teeth in the morning. Those oral amebas and flagellates that are not crushed to death by the teeth or overwhelmed by the rapid changes that take place in their environment during the consumption of three meals a day and a midnight snack, and are not swallowed and destroyed by the gastric juice in the stomach, gorge themselves with leucocytes and bacteria until they reach their maximum size and then proceed to divide into two.

One problem that offers no difficulty to these protozoa is that of transmission. Apparently human beings possess very little resistance either to infection or to the process of transmission, since about 4 out of 5 of the general population are parasitized by them. Obviously those who are not infected have never been kissed. Human nature being what it is, there is no known method of preventing the transmission of these species. Fortunately no injury results from the infection hence we don't need to lose any sleep worrying about it.

RESISTANCE OF PROTOZOAN CYSTS

All other protozoa of the human digestive tract enter the mouth in contaminated food and drink. The active stages are in most species probably destroyed before they reach their intestinal habitat, hence only the cysts are infective. Cysts are able to resist many conditions that would destroy active protozoa but are also rather delicate and succumb quickly under certain circumstances. For example, only those that are kept moist all of the time survive, since desiccation quickly destroys protozoan cysts; hence they are not blown about in dust by the wind. Many experiments have been carried out to determine the reactions of cysts to natural and laboratory conditions. Compared to Shadrach, Meshach and Abednego protozoan cysts are indeed frail, since a temperature of 113°F. for 30 minutes is sufficient to kill them; no such temperature occurs in nature except under extraordinary circumstances, therefore high temperature is not an important hazard in their lives.

On the other hand, winter has no terrors for cysts, even a winter in Medicine Hat, since no temperatures below zero to which they have thus far been subjected will kill them. Cysts will live contentedly in excrement outside of the body, or in water, for days and even weeks, which gives them plenty of time in which to find a new host. Various common disinfectants will kill them, such as creosote, carbolic acid and lysol, but seldom, if ever, are agents of this type encountered in nature. Thus the problem of the cyst is not to remain in a viable condition, but to find its way into the food or drink of a susceptible host. In most cases the cyst is able to do this because of the unsanitary conditions under which we live.

How unsanitary we really are is indicated by the fact that fully one-half of the inhabitants in many localities even in our northern states are infected with one or more species of intestinal protozoa. Only one conclusion is possible and that is that our food and drink are grossly contaminated. This is not a very pleasant conclusion to contemplate, but is one of the facts of life that should be made known to a too complacent

FIG. 80 A. Even the cook is surprised to find our food and drink so grossly contaminated with protozoan parasites.

public. Two types of control are suggested: first, efforts directed toward personal hygiene, and second, public health measures. We can make sure that food handlers in our restaurants and homes are not infected, that our water and milk are pure, that the raw vegetables we eat are properly washed and that toilet facilities are of the approved type. Communities should provide an efficient water supply, should supervise dairies, markets and restaurants, and should provide a modern system for the disposal of sewage.

LIFE CYCLES OF INTESTINAL PROTOZOA

The life cycles of most of the intestinal protozoa are similar. The cysts enter the mouth; are swallowed; and pass through the stomach unharmed. Those of the coccidium, *Isospora hominis,* hatch in the small intestine and the young and active sporozoites that emerge hasten to penetrate the cells of the intestinal wall. Those of the flagellate, *Giardia lamblia,* also hatch in the small intestine where they immediately attach

FIG. 81. Old man Balantidium undertakes a perilous journey through the mouth, oesophagus, stomach and twenty-three feet of small intestine in order to reach the colon.

themselves to the wall with the sucking disc which nature has so carefully supplied. The rest of the intestinal protozoa are carried through the entire twenty-three feet of small intestine and squeeze past the sphincter muscles of the ileocolic valve and thus into the colon, safe at last and justly proud to have accomplished successfully such a perilous journey. Many cysts hatch in the small intestine, others probably break open after they have reached the colon. The newly emerged young grow and reproduce rapidly. Many of these encyst and are carried out of the body, but enough are left to assure a continuance of the infection. How long an infection may last no one knows but probably for several years and even for life, unless proper treatment is taken.

TRICHOMONAS

"No rule is so general, which admits not some exception," otherwise we could state that all protozoa living in the intestine of man are transmitted in the cyst stage. But among the intestinal flagellates is one of particular interest because it manages to pass from one host to another in the active state, no cysts ever having been found. This species, known as *Trichomonas hominis*, is accused of causing flagellate diarrhea. The peripheral layer of protoplasm of this species seems to be thinner and less rigid than that of the other flagellates associated with it; nevertheless, this species is sufficiently resistant to withstand conditions outside of the body long enough to reach new hosts. Certainly, the weak shall inherit the living world, for as noted in Chapter III, more species of wild animals are infected with trichomonad flagellates than with any other type of protozoon. The human intestinal trichomonads will live for several weeks at room temperature in material kept in a glass container. They lived for at least six days when deposited on the ground during a damp, cool period in the spring.

For a number of years I taught my classes in protozoology that intestinal trichomonads are distributed in drinking water. Then I discovered that if they are placed in water they soon swell up like a balloon as a result of the lowering of the osmotic pressure, and die in a short time. These flagellates will, however, live for a long time in milk and presumably in various kinds of uncooked food, and in food after it is cooked and has become cool.

THE DYSENTERY AMEBA

The most notorious of the intestinal protozoa is the dysentery ameba. This species is not always satisfied to be a scavenger but occasionally becomes a blood-thirsty maneater. A large proportion of those human beings who are infected are in the so-called carrier condition; that is, the amebas live in the cavity of the large intestine, but apparently do not injure the host in any way. Host and parasite during the carrier period are really carrying on one of those wars that are not declared and not even advertized. Such a host, however, is a real menace to public health, since large numbers of amebas encyst within him, and as many as 300,000,000 cysts may be discharged by a single carrier in a single day. If we multiply this number by 10 per cent of the members of the community who are carriers, we obtain some idea of the immense numbers of infective cysts let loose upon the earth every day. No wonder our food and drink become contaminated.

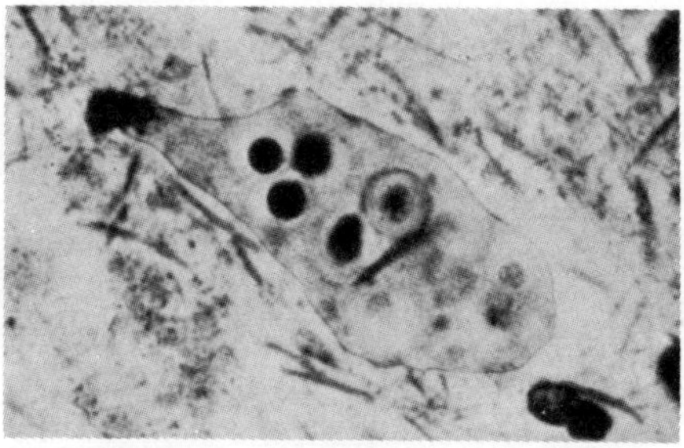

FIG. 82. This dysentery ameba is plowing its way through the contents of the intestine. The spherical black bodies are red blood cells, and the ring with a dark center is the nucleus.

Several of the most conspicuous successes in the field of preventive medicine have resulted from campaigns against diseases due to pathogenic protozoa. Perhaps the most widely known example is the control of malaria in Panama (see Chapter X). However, a number of other diseases for which protozoa are responsible, that were once a scourge in various parts of the world, can now be suppressed or eliminated entirely if our present knowledge regarding them can be put into practice. Among these is amebic dysentery.

AMEBIC DYSENTERY

This disease is not very pleasant even under the best of conditions, but before the year 1912 it was a terrible affliction.

Since that year methods of treatment have been available which prove to be effective if the disease is diagnosed sufficiently early in its course. As already noted, only a small percentage of persons who are infected exhibit severe symptoms, the majority harbor the amebas in their large intestine, but suffer little or no disturbance as a result of their presence. When the host is especially susceptible, or the parasite un-

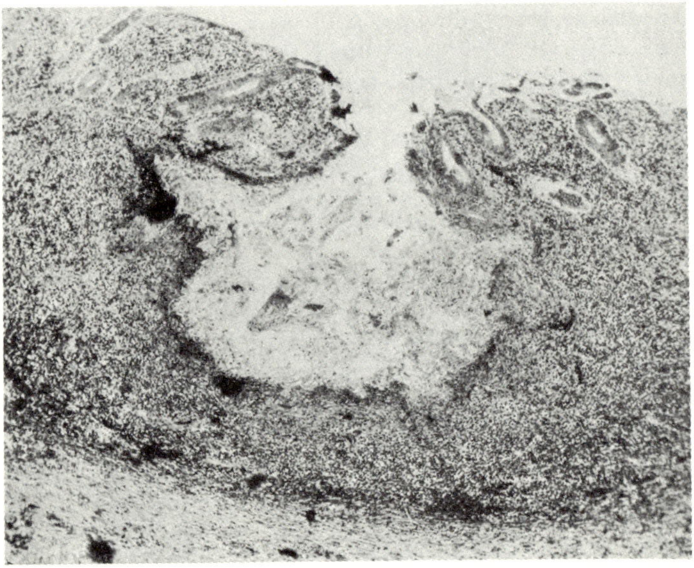

FIG. 83. This is a section of the large intestine of man which shows what an amebic ulcer looks like under the microscope.

usually virulent, the amebas succeed in penetrating the intestinal wall. Here they live on tissues that are dissolved by enzymes secreted by themselves. Such an invasion results in the formation of an abscess, which may break out into the lumen of the intestine, thus forming an ulcer. The symptoms accompanying such an invasion are at first diarrhea and later dysentery. The host may recover spontaneously or secure relief with drugs, otherwise the invasion continues, and the symptoms increase in severity. Eventually death ensues from exhaustion, intestinal hemorrhage or peritonitis following the perforation of the intestinal wall. Often the symptoms simulate appendicitis or gastric ulceration and mislead physicians, who are not well acquainted with amebiasis, to perform operations.

Before satisfactory drugs were elaborated for the treatment of amebiasis, death was not uncommon, especially in those who ventured to live in the tropics where this disease has always been more prevalent and more virulent than in temperate regions. During the past two decades a number of

FIG. 84. Often the protozoologist gives the guinea-pig a vacation and tries a few experiments on himself—but not for long.

expeditions have been organized in my department for the purpose of studying parasitism in the tropics, and in every case, in spite of all precautions, one or more members became infected with dysentery amebas, and several members came down with active symptoms of amebic dysentery. There have been no serious consequences, however, since we knew how to make early diagnoses and how to treat the disease. As a matter of fact, several of us elected to remain infected after we returned to our laboratories so that we might serve as guinea-pigs for testing the effects of certain drugs on the parasites.

AMEBIC ABSCESSES

Amebic dysentery is only one manifestation of amebiasis. Organisms that invade the intestinal wall may gain entrance to the capillaries and be carried to various parts of the body in the blood stream. Since the entire blood supply passes from the heart throughout the body and back again to the heart in from twenty to thirty seconds, the amebas are well provided with transportation. Wherever conditions in the body are favorable the amebas may set up secondary infections. Most frequently these secondary infections occur in the liver, where the amebas dissolve the tissues, just as they do in the intestinal wall, and bring about the formation of abscesses some of which may be larger than a baseball. Amebic abscesses may also be formed in the lungs, brain, spleen and other parts of the body. Brain abscesses terminate fatally, but liver abscesses may be alleviated by means of drugs or by surgical operation. On the whole the dysentery amebas treat us pretty well, since they rarely attack us seriously, although they could riddle us with ulcers whenever they wished to do so.

TREATMENT OF AMEBIASIS

Few people realize how much modern medicine owes to primitive man. From him we have inherited drugs effective against both malaria and amebiasis. Early South American explorers discovered that the natives used a decoction of the root of a plant for the treatment of diarrheas and dysenteries. The dried roof of this plant (*Psychotria ipecacuanha*), was known as ipecacuanha. It was introduced into Europe in 1658 and was used in the treatment of dysentery in India in 1660. An immeasurable amount of suffering and hundreds of thousands of deaths could have been prevented if ipecacuanha had been employed consistently from that time on, but for some unknown reason its therapeutic value seems to have been buried in oblivion for over a century. Then, in 1817 Pelletier extracted the alkaloid known as emetine from the ipecacuanha root, and in 1829, Bardsley applied the extract with success to patients suffering from diarrhea and dysentery, but it was not until 1912 that its use in the treatment of amebiasis was generally adopted.

In that year, Vedder, a member of the medical corps of the United States Army, demonstrated that ipecacuanha from which emetine had been extracted possessed very little amebicidal action, but that emetine itself when used alone destroyed free-living amebas in a dilution of 1 to 100,000. On the basis of these results Sir Leonard Rogers immediately introduced emetine into India for the treatment of amebiasis with such success that the drug was soon employed throughout the world and amebic dysentery at once lost its terrors. From that time on life in the tropics has not been menaced continually by this dread disease. More effective drugs than emetine are now known, such as an arsenical preparation called carbarsone and a compound containing a large percentage of iodine which is sold under the trade names chiniofon, yatren or anayodin.

As in most protozoan infections, a drug may eliminate all of the parasites in many cases, but in other cases a few of the organisms seem to be resistant. Apparently amebas that are resistant to one drug may not be resistant to another drug, hence, if they are not all removed by the first course of treatment, a second or third drug can be resorted to. Excellent results have been obtained in this way.

Drug prophylaxis for the prevention of infection with dysentery amebas has never been practiced to my knowledge. We have formulated plans for carrying on experiments of this nature but have not yet had an opportunity to test them. There seems to be no reason, however, why one who lives in a region where amebiasis is prevalent cannot destroy the organisms that may find their way into the intestine by taking doses of an amebacidal drug at intervals of a few days, just as doses of quinine are administered to those who visit a malarious region. Arsenicals, such as carbarsone, are unfavorable for such a purpose since their toxicity to the human body is cumulative, but chiniofon exhibits no such reaction and can be employed without the necessity of modifying one's daily activities.

EPIDEMICS OF AMEBIASIS

Amebiasis is endemic throughout the world, that is, it is present at all times among the general population. Epidemics are not common but do occur from time to time. Thus at the time of the Philippine Insurrection epidemics of amebic dysentery broke out among the United States troops. Again in 1916, United States soldiers stationed in the El Paso, Texas, district suffered a distinct epidemic of this disease. No one, not even a United States Marine, is tough enough to withstand the rapid fire guns of the dysentery amebas. How many wars they have won cannot even be estimated, but they have coöperated with malaria organisms and other protozoa and with typhoid, typhus, etc. in shaping the destinies of nations throughout the centuries.

The most conspicuous epidemic of amebic dysentery known to have occurred in a civilian population was the Chicago epidemic of 1933. The residents of Chicago are ordinarily no more highly infected with dysentery amebas than those of other cities in the temperate zone, but on August 15, 1933 a number of cases of amebic dysentery occurred in two Chicago hotels. These were at first supposed to be due to the contamination of food by infected foodhandlers, hence 364 foodhandlers were examined and 7 per cent of them were found to be carriers. A month later the 338 foodhandlers who were negative on the first examination were reëxamined and 18 per cent were recorded as positive for dysentery amebas.

Evidently the incidence of the infection was increasing. At the same time 498 non-foodhandlers were tested and 20 per cent of them found to be positive. Some of the active cases that developed in the meantime appeared in guests who had remained at the hotels for only a short time. The incubation period, that is the interval between the entrance of the parasite into the digestive tract and the appearance of symptoms, was notably short in these cases and the symptoms exceptionally severe. These infections were such as result when very large numbers of cysts are swallowed in a brief period, hence they were evidently not due to the contamination of food by infected foodhandlers since only a few cysts could spread in this way.

A search was made to see if the water supply was satisfactory. Soon a leak was discovered from a sewer pipe into a tank of ice water. This water was being distributed throughout the hotels. After this leak had been repaired no further cases of amebic dysentery appeared. Evidently cysts discharged into the sewage system accumulated in such numbers in the ice water that was used for drinking purposes that massive doses were ingested by both servants and guests. Cysts will probably live longer in ice water than under any other conditions, hence the situation was ideal for their viability.

Questionnaires were sent out to all who had been guests of these two hotels and the answers received compiled. The data indicate that about one thousand cases of amebic dysentery were contracted in these hotels resulting in fifty-eight deaths. Guests on leaving the city traveled to all parts of the country before symptoms appeared and cases were reported from almost every state in the union. Doctors in certain localities were unfamiliar with the disease hence proper treatment was withheld until too late and deaths ensued that might have been prevented.

The Chicago epidemic created a real furor among physicians, and professional men, and a considerable proportion of the general population became, for the first time, ameba-conscious. One important result of the epidemic was the overwhelming evidence that ameba cysts may be spread in drinking water. Whether the alcoholic content of beer, wine and hard liquor is high enough to pickle the infective cysts is a problem that offers some very attractive experimental prospects.

PROTECTION FROM INTESTINAL PROTOZOA

Amebic dysentery is usually listed among the tropical diseases but it is now evident to everyone that we must also guard against the disease in temperate regions. This applies to infections with other intestinal protozoa as well as with dysentery amebas. An important question is what to do after one has become infected. Three courses are open to us. We may help the body in its fight against the invading parasites by resting as much as possible, thus conserving instead of dissipating energy needed in building up resistance. We can treat certain of the infections with drugs; but for several types of protozoan parasites no effective drugs are available. And we can modify our diet in such a way as to render the contents of the intestine, where the invaders live, unsuitable for the growth and reproduction of the protozoa. The

FIG. 85. Every protozoon found guilty of injuring the human body is considered a public enemy and relentlessly pursued by the protozoological G-men.

plumper of my readers will be glad to learn that the best diet thus far elaborated is of the reducing type.

How to change the environment within the intestine so as to make it unfavorable for the parasites is an interesting and important subject. A number of investigations have been carried on in my laboratory that have a bearing on this problem. These were begun because of the scientific aspects of the work, but any clues that turned up that might lead to the control of the parasites were followed with as much enthusiasm as the G-men pursue our Public Enemy No. 1. The story of these researches effectively demonstrates once more the way in which the results of study in the field of pure science may at any time become of practical significance.

OPALINID CILIATES OF FROGS

Certain ciliates, known as opalinids, are commonly present in the rectum of frogs and toads. The green frog is a notable exception, since no opalinids are present in adult green frogs, although the tadpoles of green frogs are frequently infected. It had been suggested that, since frog tadpoles live on vegetation and young frogs, after they leave the water, live on insects and worms, perhaps the change in diet at this time might be responsible for the disappearance of the opalinids.

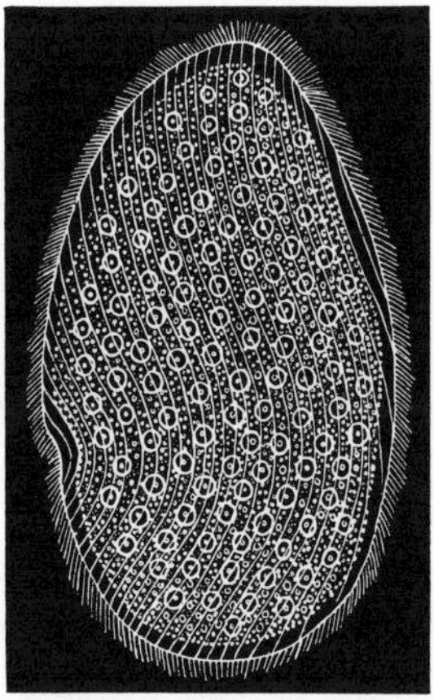

FIG. 86. The opalina that lives in the rectum of the frog has many nuclei but no mouth. It absorbs nourishment through the body wall.

Following up this suggestion, I examined a large number of tadpoles of the green frog in various stages of development and found that opalinids were numerous in young tadpoles. During metamorphosis, when the legs appear and the tail is adsorbed, changes also take place in the morphology of the alimentary canal and the intestine decreases in length to about one-fourth of that of the young tadpole.

The opalinids must become quite distracted by these catastrophic changes; some of them like homeless refugees migrate forward from the rectum into the intestine; later all of them disappear from the rectum and finally those in the intestine also capitulate to nature's caprices. In other species such as the common leopard frog, opalinids remain in the rectum throughout the entire period of metamorphosis and into the adult stage. Since the leopard frog and green frog both feed largely on insects and worms after metamorphosis, some factor other than the character of the food must be responsible for the loss of opalinids in the adult green frog. The opalinids disappear when the intestine and rectum become shorter, but this shortening also occurs in the leopard frog, and in other frogs, without any effect upon the ciliates. Thus we have certain observations, but no real clue as to what happens to the opalinids of the green frog. However, our attention was definitely directed by these investigations toward the effects of dietary changes on intestinal protozoa.

PROTOZOA OF CARNIVOROUS ANIMALS

Soon an opportunity presented itself to study a group of rats that had been fed on various types of diets. No conspicuous differences were noted among the rats with respect to their protozoan inhabitants except in one lot of three that were entirely free from intestinal forms except for a few specimens of a small flagellate known as *Hexamita*. Rats are usually highly infected with protozoa; trypanosomes live in the blood, giardia flagellates and coccidia in the small intestine, and so many species and individuals of amebas and flagellates usually occupy the cecum that its contents often seem to consist of nothing but a squirming mass of tiny wrigglers. When the dietary history of these three experimental rats was examined we found that they had been fed throughout their lives on a diet consisting largely of meat and that their mothers before them had been maintained on a carnivorous diet.

This discovery was naturally very stimulating and as a result the astonishing fact occurred to me that practically all of the intestinal protozoa we were using in our classes came from vegetable-eating animals. These included protozoa from the stomach of cattle, from the cecum of the rat, guinea-pig, and chicken, from the colon of the pig and even from the professors and instructors in the department. A hasty survey of the literature also revealed a surprising situation; very few intestinal protozoa have ever been described from carnivorous animals. A group of carnivores maintained in one of the Baltimore parks were then examined and the paucity of intestinal protozoa in meat-eaters again demonstrated.

FIG. 87. Three of the flagellates that live in the intestine of rats are *Giardia muris* (1), *Hexamita muris* (2), and *Trichomonas muris* (3).

EFFECTS OF A CARNIVOROUS DIET ON RATS

By this time we had organized an experiment to test the effects of a meat diet on intestinal protozoa. Two groups of rats were examined and all found to be infected with trichomonad flagellates. One group of 20 rats were fed for two weeks on what we call a normal diet consisting largely of carbohydrates and the other group of 20 rats were fed for two weeks on an experimental diet made up principally of animal proteins. At the end of the two weeks the rats were anesthetized and examined. The control rats contained enormous numbers of trichomonads; whereas the experimental rats contained very few or none at all. The common giardia flagellates of the small intestine appeared to be similarly affected, but a smaller species, *Hexamita muris*, was as abundant as in the controls. Evidently the changes in the intestine due to the carnivorous diet made life unbearable for the trichomonads, and at least unpleasant for the giardias, but the hexamitas seemed to thrive on good red meat.

Special attention was then directed toward the trichomonads for several reasons. In the first place these flagellates are widely distributed among animals in nature and are almost always present in laboratory rats. Second, they can be cultivated with the greatest of ease; all that is necessary is to add a little dried and powdered blood to saline solution, put it in a test tube and maintain it at a temperature of 37°C. And third, trichomonads seemed to be more profoundly affected by the change brought about by the meat diet than were the other species of protozoa.

EFFECTS OF A CARNIVOROUS DIET ON MAN

Having demonstrated the detrimental influence of a meat diet on trichomonads in rats, the next step was to try this treatment on human beings suffering from trichomoniasis. Members of the department of biochemistry who were interested in nutrition made up a series of menus which included items consisting largely of animal protein but contained the elements necessary for satisfactory maintenance. Almost immediately several patients suffering from flagellate diarrhea were located and the meat diet tried with amazing results. How simple the life of a scientist would be if he could only don the gloves and settle disputed questions with a boxing match. As a result of my experiences, I should certainly bet on the meat eater.

The first human patient we experimented on had suffered intermittently for twenty years and was so emaciated as to fit very well the expression "living skeleton." After four days his symptoms disappeared as well as the flagellates and at the end of a week he insisted on returning to his home in Georgia. Whether or not he had a relapse I do not know. Other patients recovered with equal rapidity although very skeptical at first. Included in our menus were a number of items, such as fruits and relishes, designed to make the diet as palatable as possible. One of the patients who didn't understand the rationale of the treatment was very much disturbed because she couldn't get any strawberries and another patient had a terrible time because she "just couldn't eat pickles." Otherwise the treatment was a success.

FIG. 88. The uninfected meat eater knocks out the vegetarian who is full of parasites; wouldn't it be lovely if we could settle scientific problems so easily?

Flagellate diarrhea is not encountered very frequently in temperate regions, but the patient who suffers from it feels that he, at least, is 100 per cent infected. Our scientific interest in the environmental control of intestinal protozoa has not only given such patients relief, but has also removed one more peril from life in tropical countries where intestinal disturbances are prevalent and trichomonad flagellates abundant.

HYPOTHESES REGARDING EFFECTS OF CARNIVOROUS DIET

The most conspicuous modification of the intestinal contents due to a meat diet is the development of a preponderance of the putrefactive type of bacteria. For example, the intestinal bacteria of a rat living on what we call a normal diet, consisting largely of carbohydrates, are present in the proportion of ninety-nine of the aciduric type to one of the putrefactive type. If such a rat is fed for a few days on a diet high in animal proteins a complete reversal in the proportions of these two types of bacteria occurs; that is, a change to ninety-nine of the putrefactive type to one of the aciduric type takes place. Bacteria might affect the trichomonad flagellates in several ways. First, the flagellates are known to ingest bacteria and use them as food; perhaps they have such a violent idiosyncrasy against the putrefactive type that they starve to death rather than eat them. Bacteria of this type probably represent to these flagellates the spinach of the microscopic world. Second, the decomposition products of the two types of bacteria are distinctly different. The putrefactive type give rise to skatol, indol, hydrogen sulphide, putresins, cadaverins, and many other kinds of offensive substances. No wonder the protozoa can't take it. Acidurics merely produce lactic, acetic and other harmless acids. And third, the hydrogen-ion concentration of the intestinal contents may be so modified by the putrefactive bacteria that life becomes impossible for the poor little flagellates.

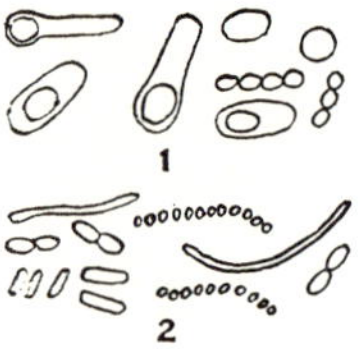

Types of bacteria. 1. Proteolytic anaerobes. 2. Acidurics.

The third hypothesis was tested first, and the change in the hydrogen-ion concentration was found to be so slight as to be of no significance. Later work confirmed this conclusion. Then a careful study of the bacteria was begun. Briefly it was found that the number of trichomonad flagellates decreased as the number of proteolytic anaerobes increased and increased as the number of proteolytic anaerobes decreased. An anaerobe is a type of bacterium that lives without air or free oxygen and a proteolytic anaerobe is one that can dissolve proteins.

EFFECTS OF DIFFERENT TYPES OF PROTEIN

The next series of experiments was designed to determine the efficacy of different types of protein diets. Three types of diets were prepared, each containing 60 per cent protein and other elements necessary for satisfactory growth and reproduction. On a diet consisting largely of beef liver, the proteolytic anaerobes did not become more numerous, but the trichomonads increased in numbers. On a diet of beef steak, an increase in proteolytic anaerobes was noted and a decrease in acidurics and in the number of trichomonads. On a casein diet, a similar but more marked increase in proteolytic anaerobes occurred and a greater decrease in acidurics and in trichomonads was observed. One of the most painful features of these experiments was that of feeding good beefsteak at sixty cents per pound to rats, while our meagre salaries made it necessary for us to live on Philadelphia scrapple and on an occasional laboratory animal, such as a chicken or a rabbit, that had served its purpose as an object of experiment. I even suspect that one of my associates actually selected chickens for his investigations so that he could later eat up the evidence.

These experiments confirmed the previous conclusion that the decrease in the numbers of trichomonads is due to an increase in proteolytic anaerobes; they also contributed another interesting result, namely, proteins differ with respect to their influence on the development of proteolytic anaerobes, hence the character of the protein in the diet must be controlled if the proper bacteria are to be encouraged and the trichomonads eliminated. According to these experiments a liver diet, which has been found to be so wonderfully efficacious in cases of pernicious anaemia, has no obvious effect on trichomonad infections. Casein appears to be the best animal protein yet employed against these flagellates, therefore we now recommend that patients eat as much cottage cheese as possible, since cottage cheese consists entirely of casein.

If we had human beings to experiment on our work would no doubt progress much more rapidly. I have suggested that some of our less desirable professors might volunteer for this purpose, but thus far none of them has been willing to sacrifice himself for the cause of science, even though I have given my promise that if anything went wrong I would guarantee them a lovely funeral with only protozoologists as pall bearers. Consequently we are still using rats.

INFECTION OF STARCH-FED RATS WITH BALANTIDIAL CILIATES

The next chapter of this story concerns another type of intestinal protozoa, the large ciliate, *Balantidium coli*. This species is not common in man except in certain favorable localities. It is usually present in pigs and often in chimpanzees. Several investigators had stated that they could determine whether a pig was or was not infected with this species by observing the quantity of undigested food residue, including

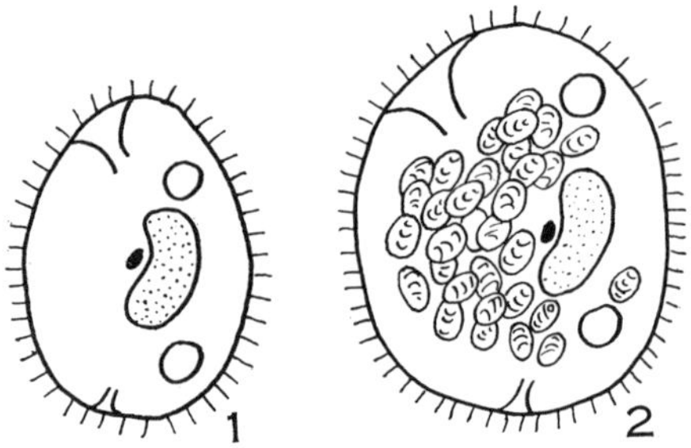

FIG. 89. *Balantidium coli* (1) before and (2) after gorging itself with starch grains.

starch granules, in the intestinal contents. The greater the amount of starch the larger the number of balantidia. We not only found this to be true in Baltimore pigs, but also were able to prove that heavy infections were accompanied by large numbers of aciduric bacteria, and light infections with a preponderance of proteolytic anaerobes, thus confirming our previous results with the trichomonads in rats.

Rats in nature are not normally infected with balantidia, and rats that are being fed on a "normal" diet cannot readily be infected with them; but rats that were fed on a diet high in carbohydrates were infected with balantidia without difficulty and retained the infection for as long as five months.

When infected rats were placed on a "normal" diet they lost their balantidia within a few days. When rats were fed on a diet containing as much as 70 per cent casein they could not be infected and those that were already infected soon lost their balantidia.

One of the most important carbohydrates is starch. For this, and several other reasons, starch was used in our next experiments. About ten years ago someone discovered that the addition of starch grains to culture media resulted in a better growth of dysentery amebas. Since then practically everyone who has cultivated amebas in test tubes has found starch to be of great value in promoting growth and multiplication. We have found that starch is favorable for the development of trichomonads and balantidia in cultures. The starch grains are ingested by the protozoa and apparently digested and used for nutritive purposes. However, not until we placed hungry balantidia under the microscope and then added starch to the medium did we realize how voracious a protozoon can be. The balantidia did not ingest the starch grains one by one and at infrequent intervals, but proceeded to gorge themselves with a dozen grains at a time so that within a few minutes they were loaded with starch and had increased in bulk 100 per cent or more. The gorged appearance of these balantidia suggested that we name them after certain of our more dignified colleagues. Then, for a while, we had the unique pleasure of observing these solemn professors swimming about under our microscope just as nature, and a healthy appetite, had made them.

RELATION BETWEEN STARCH AND INTESTINAL PROTOZOA

These observations led us to test out a new hypothesis, namely, that intestinal protozoa depend on starch for their food supply and when a high protein diet is fed to the host the organisms starve to death because of the lack of starch. It is well known that protozoa, unlike human beings, practice very rigid food selection; they will eat one or several types of food but prefer to die rather than eat anything else. Our experience both with starch in cultures and with carbohydrates in the diet of rat and human hosts favored the hypothesis that intestinal protozoa prefer starch as an item of food. To test this further we attempted to determine whether a correlation exists in man between the amount of starch in the intestine and the number of protozoa present. To do this it was necessary to study a group of people that were accustomed to eat large quantities of starch. Accordingly funds were obtained for a

FIG. 90. We named some of the voracious balantidia after certain of our more dignified colleagues.

survey to be carried out in Central America where the natives live largely on corn cakes (tortillas), rice and beans.

But first we had to learn how to extract starch from the intestinal contents. A large part of one winter was spent on this difficult problem but we finally devised a satisfactory technique.

We selected Guatemala as the scene of our hunting expedition and spent a more or less pleasant summer among the natives of that Republic. Contrary to general belief many tropical countries possess a delightful climate during the summer, which is the only season of the year when many of us can get away from our teaching duties for any considerable length of time. The principal problem is that of securing the necessary funds for expeditions. Several years ago, after I had with considerable difficulty accumulated funds enough to take a long coveted trip to Honduras, the wife of one of my colleagues, who is not a parasitologist, expressed great concern regarding my comfort and health and asked why I didn't get one of my assistants "to go to that awful country." After all, the study of parasitic organisms is sometimes very dangerous. How many martyrs to science are numbered among those who have tried to solve some of the problems presented by pathogenic species is not known but the hunter is often successfully hunted. Fortunately we now possess methods of creating immunity to the invasion of certain species and therapeutic drugs are known in case an investigator becomes infected by other species.

Guatemala was chosen for our work because the natives were known to live largely on carbohydrates and to be highly infected with intestinal protozoa. The typical native family in Guatemala lives in a flimsy hut with walls of upright poles, a thatched roof, dirt floor and no toilet facilities. Pigs, chickens, ducks, dogs, cats, the cow and burros all use the same patio with the family and not infrequently the animals must be chased out of the hut so they will not wake up the baby. Add to this myriads of flies and fleas by day, and rats, mice, mosquitoes, cockroaches and bedbugs by night, and you get a fair idea of nature in the raw. But what a treat for a naturalist! However, everyone sleeps in a hammock over 8 inches above the floor and hence escapes the fleas by night as well as the invasion of some of the other types of vermin. Breakfast, dinner and supper consist largely of rice and beans with meat perhaps once a week and salt codfish occasionally. Bananas are of course a staple. Food is cooked over a charcoal stove made of clay and the water supply for both drinking and cooking is so enriched with organic matter that it both "tastes and smells."

Under these conditions the poor natives have about as much chance to escape malaria and infections with intestinal protozoa as has the proverbial snowball. Visiting naturalists are regarded by the children as playthings sent to them by a kind and loving Father and by their parents as harmless lunatics running about with a butterfly net and breaking out with rashes of ecstasy over the discovery of species of insects that are new to them.

The most important phase of a foreigner's life in the tropics is that of protecting himself from the attacks of parasites. The natives are truly amazed when you demand boiled water to drink and insist on seeing it boiled. The entire family gathers around while you immerse your fresh vegetables in boiling water for 10 seconds to kill the protozoan cysts. They probably do not realize that everything else you eat is either thoroughly cooked, comes out of a can, or has a covering, such as the shell of an egg or the skin of an orange, that protects the substance within from both protozoa and bacteria. Many travelers carry with them some type of disinfectant for sterilizing water when boiling is not convenient. These disinfectants, however, are deceptive since they do not kill protozoan cysts and hence do not protect one from amebiasis, giardiasis and all the other iases due to intestinal protozoa.

During a year spent in the Philippines, we had an opportunity to visit the so-called Moros and the Sultan of Sulu on the island of Jolo. How to obtain safe drinking water was the principal problem involved. We decided to use boiled water whenever possible, and to take a disinfectant along to use in emergencies, since this would protect us at least from bacterial infections. We purchased a demijohn of a disinfectant with which we started the trip, but failed to read the directions. When the time came to use the drug we found that two drops per gallon of water was the amount advocated. According to this we had almost enough of the chemical with us to disinfect the entire Pacific ocean. At any rate we escaped infection.

A hospital is an excellent source of material for the study of intestinal protozoa since the director, resident physicians and nurses are sympathetic toward any type of research that promises to relieve human suffering, and hospitals can usually furnish a laboratory and also obtain material from patients with little difficulty. On this account, we established ourselves in Guatemala in the United Fruit Co. hospital at Quirigua. The samples secured for our use were first thoroughly mixed with saline solution (0.7 per cent NaCl). Saline solution is used since water might lower the osmotic pressure to such an extent as to destroy any protozoa that chanced to be present. A drop of the mixture was then placed on a slide and spread out under a coverglass. If moving protozoa were detected under low power they were studied under higher magnification so as to make their identification certain. Protozoan cysts were likewise examined highly magnified. Certain of these are easily identified by their shape and size, but others, especially those of certain species of amebas, must be treated with an iodine solution, which stains the nuclei, before the species can be determined. In a few cases, even this treatment is not sufficient for an accurate diagnosis and the iron-hematoxylin method of staining has to be resorted to. This method is so complicated that it would be confusing to attempt to describe it here.

After examining the specimens available at this hospital we proceeded to the town of Chiquimula. The Friend's Mission kindly provided quarters for us and also furnished a supply of samples mostly from children attending their private school. The missionaries here had become famous among the physicians of Central America because of the large number of cases of amebic dysentery among them. As a result of our studies we obtained information that indicated rather definitely the importance of drinking water as a method of distributing the infective cysts responsible for this disease. The incidence of infection with dysentery amebas had fallen off remarkably since a new and uncontaminated water supply had been installed eighteen months previous to our visit. Health officers are constantly debating the rôle that drinking water plays in the transmission of protozoan cysts; on this account every observation of this type is of value in the solution of this problem.

Our next work was done in Guatemala City where material was obtained in part from the American Hospital and in part from inmates of the federal penitentiary through the efforts of the department of health.

FIG. 91. Like Gulliver man is held in bondage by his Lilliputian parasites

On our return to Baltimore we examined the samples we had brought back with us; this was done with very high magnification so as to bring out the minute nuclear details necessary for definite diagnosis. One hundred samples were examined in this way so as to obtain accurate data regarding the species of protozoa contained in them and the comparative number of individuals of each species. Then 100 cc. of each sample, that had been preserved in absolute alcohol, were analyzed to determine the quantity of starch present. This work occupied a large part of our available time during the following winter. I wish I could tell you that we were successful and that our hypothesis was proved to be correct. On the contrary it is necessary for me to report that our results gave no indication whatever of a correlation between the quantity of starch and the growth and multiplication of intestinal protozoa within the human host.

The details of most of these experiments have been published from time to time during the past fifteen years in language intelligible only to specialists and with many tables of data treated by statistical methods. It would obviously be cruel and unusual punishment to inflict these details on the general public. Perhaps I should not have written about the unsuccessful outcome of our latest investigations, but this may be considered an example of the discouragements that are more often than not the outcome of research work. At any rate you have been introduced to a scientific borderland that some of us consider to be of great importance in its relation to human welfare.

In spite of our failure we still believe that the hypothesis we attempted to prove in Guatemala is a good one and are waiting for some inspiration to suggest a practical method of testing it. We hope that we may be granted the pleasure of proving it to be correct; but whether it is correct or not is after all of little consequence since we are searching for the truth, and in this case there is no doubt that the truth will set us free from the bondage imposed upon us by our Lilliputian intestinal protozoa.

Chapter VIII

OH, EAST IS EAST AND WEST IS WEST

AMONG the blessings enjoyed by the inhabitants of the United States is freedom from certain Tropical Diseases that might really give us something worth worrying about. Among these diseases are African sleeping sickness and Chagas disease, or Trypanosomiasis, and oriental

FIG. 92. The trypanosomes responsible for African sleeping sickness and Chagas disease have thus far been excluded from the United States.

sore, espundia and kala azar, or Leishmaniasis. Anyone who has seen a human being suffering from either of these afflications may well be thankful that his ancestors came over in the Mayflower, or at least reached this country before the immigration quota was completed. That we should not be too complacent regarding these diseases, I hope to demonstrate before reaching the end of this chapter.

TRYPANOSOMES

The term trypanosomiasis is applicable to any infection with flagellated protozoa belonging to the genus *Trypanosoma*. The trypanosomes are practically 100 per cent aquatic, and might almost be considered marine animals, since most of them spend a considerable proportion of their lives in the blood of warm-blooded animals which contains as much as 0.6 per cent of sodium chloride (blood plasma of man). Usually it is a simple matter to secure trypanosomes for study since many frogs and newts are infected with them. All that is necessary is to prick a blood vessel of one of these animals,

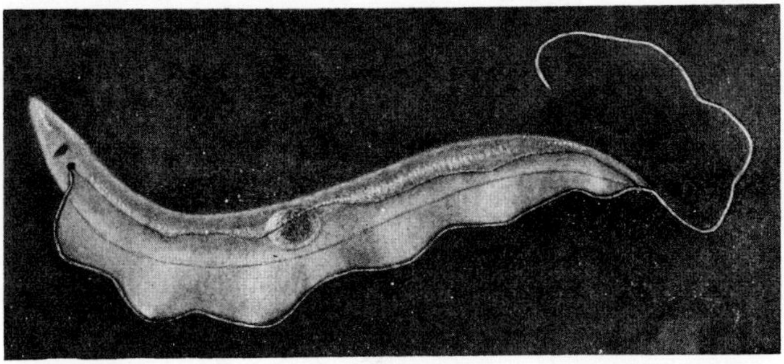

FIG. 93. A typical trypanosome with parabasal body at the posterior end, undulating membrane along the side and flagellum extending out from the anterior end.

place a drop of blood on a slide and examine it with a compound microscope. If living trypanosomes are present groups of blood corpuscles in various parts of the field will be agitated by the violent movements of a flagellate in their midst. Trypanosomes, like Cassius, have "a lean and hungry look"; such parasites, as we shall find, "are dangerous."

But to realize the true beauty of a trypanosome, a preparation must be made by drying a blood film and staining it so as to render the body blue and the nucleus and flagellum red. Near one end is a deep red spot that looks like an eye but is only a parabasal body: in fact, this spot is close to the posterior end. The undulating membrane extends almost the entire length of the organism, and, although the flagellum that is attached to its outer margin projects freely like a tail, this is actually the anterior end. Unfortunately "handsome is as handsome does," hence it becomes my painful duty to convict these graceful, serpentine reprobates of some of the most revolting offenses known against man and animals.

FIG. 94. The best way for a parasite to gain publicity is to attack human beings.

To some of us it is a bit difficult to understand why so many people take an interest in criminals and attempt to succor them no matter how shocking their assults against society may have been. There is no difficulty, however, in explaining man's interest in criminal types of trypanosomes, since no better method of gaining publicity exists among the protozoa than that of attacking man or an animal useful to man. To biologists the life cycles and methods of transmission of the trypanosomes are so extraordinary that our interest in them would be excited even if some of them were not pathogenic.

TRYPANOSOMIASIS IN RATS

Probably more is known about the trypanosome that lives in the rat than about any other species. A rat that has become parasitized does not exhibit any symptoms and evidently doesn't know he has been invaded. Anyway he couldn't tell us how badly he felt even if he wanted to. However, we can determine his condition by examining his blood from

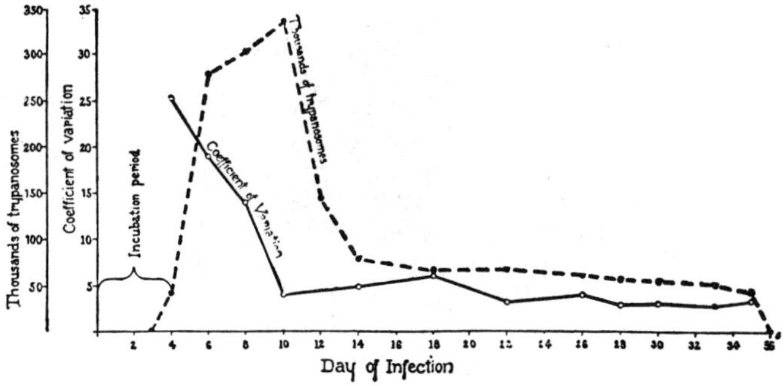

FIG. 95. These curves indicate the thousands of *Trypanosoma lewisi* per cubic millimeter of blood in a rat and their variations in size during an infection.

day to day. As the accompanying diagram shows the organisms appear in the blood about three days after infection; they multiply rapidly by longitudinal division up to about the tenth day; then they decrease in number even more rapidly until the fourteenth day after which they decrease in number very slowly, finally all disappearing in about thirty-six days.

Although the rat is not conscious of his infection, his body reacts against the presence of the trypanosomes in several interesting ways. In the first place a substance develops in the rat's blood that inhibits the reproduction of the trypanosomes so that after the tenth day division ceases entirely and only adult flagellates are present. The rate of reproduction of the trypanosomes can be determined by computing the coefficient of variation from measurements of the length of the organisms. Depriving the poor trypanosomes of the joys of reproduction ought to be punishment enough, but at about this time, another substance appears in the blood that actually destroys most of them. Apparently those that are left are all crime-hardened adults that at first gradually and then more rapidly die of old age. Probably no one envies the rat, but so far as trypanosomiasis is concerned his lot is a happy one compared

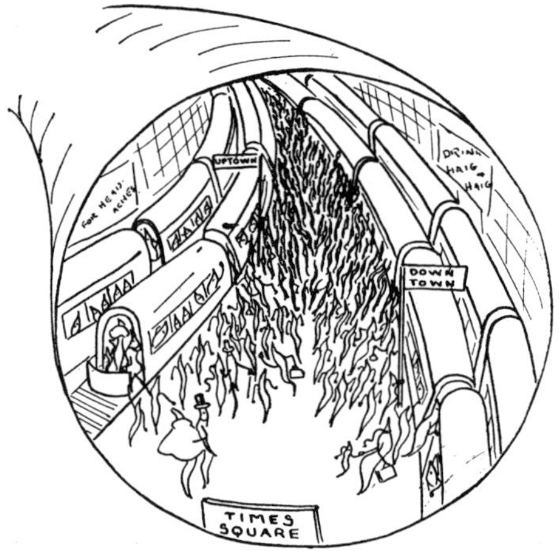

FIG. 96. A blood vessel full of trypanosomes reminds one of the subway at Times Square during the rush hour.

with that of human beings who are so luckless as to become infected.

The reproduction-inhibiting substance that appears in the rat's blood is unique in the field of immunology, since no such product has ever been discovered in any other animal as a result of protozoan infection. It can be transferred from an infected to a clean rat by withdrawing blood from an animal that has been infected for about ten days and injecting it into an uninfected rat. If blood containing trypanosomes is then inoculated into this rat no division occurs among these trypanosomes, thus demonstrating the extraordinary power of this reproduction-inhibiting substance. If such a product could be developed in man what a boon it would be to those who believe in birth control.

In contrast to infections in the rat, due to the species (*Trypanosoma lewisi*) that ordinarily occurs in this animal in nature, are infections resulting from inoculations with other species of trypanosomes. For example, if a rat is infected with a trypanosome pathogenic to horses neither the reproduction-inhibiting product nor the substance that destroys trypanosomes appear, consequently the trypanosomes are able to multiply continuously until the blood is a wriggling mass of flagellates. Eventually the trypanosomes commit suicide by killing the rat. A blood vessel of such a rat if highly enough magnified reminds one of the subway at Times Square during the rush hour.

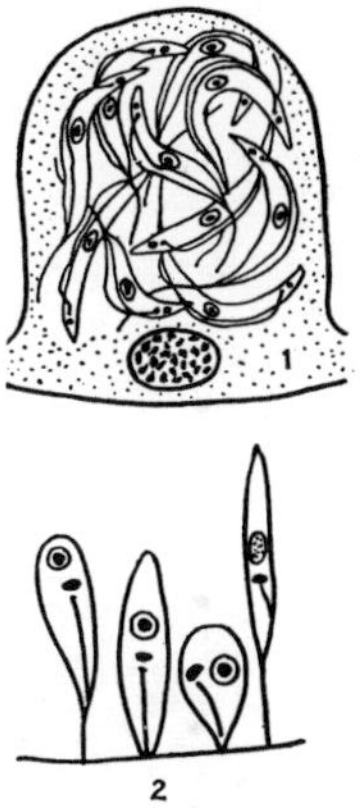

Trypanosomes of the rat in the flea, (1) within the cells of the stomach wall, and (2) attached to the wall of the rectum.

RAT TRYPANOSOMES IN THE FLEA

Even more remarkable than the curious course of the infection of the rat trypanosomes in rats is the life cycle of these flagellates in the transmitting agent, which is the rat flea. When a flea feeds on an infected rat, blood containing trypanosomes is drawn into the stomach. The trypanosomes are not digested, but hasten to penetrate the cells of the stomach wall. Here safe from the digestive juices of the flea, they grow and reproduce extremely rapidly, being restrained only by the cell walls. The young trypanosomes soon break through these walls and escape; they may then penetrate other cells of the stomach wall or be carried out of the stomach and through the intestine until they reach the rectum of the flea. Some of them succeed in anchoring themselves with their flagellum to the walls of the rectum. These have demonstrated their fitness to survive and are allowed by nature to remain there and reproduce as long as the flea lives. Such a fate may satisfy a successful trypanosome, but as for me, I shouldn't care to live in the rectum of a flea. The trypanosomes are now transmitted in one of two ways; the rat may eat an infected flea or may lick the excrement of an infected flea from its fur.

TRYPANOSOMES OF MAN

Several hundred species of trypanosomes have been described from vertebrates and no large group of animals with backbones seems to be free from them. Fortunately most of these species are limited to lower animals, only three of them being human parasites. These three, however, are all pathogenic and of primary significance to human welfare in those regions where they are prevalent. African sleeping sickness is due to one or the other of two nearly related species and American trypanosomiasis, or Chagas disease, to a third species. The life histories of the two African species are similar but differ considerably from that of the American species and from that described for the type that lives in the rat.

African sleeping sickness is not related to what we call sleeping sickness in this country. Our disease, known as encephalitis lethargica, also involves symptoms such as apathy, drowsiness and lethargy, but is due to a filterable virus and not to trypanosomes.

The trypanosomes of African sleeping sickness are transmitted by tsetse flies belonging to the genus *Glossina*. These and other species of flies may bite an infected person or antelope and then transfer trypanosomes to another person or antelope if it bites again immediately. Ordinarily, however, mechanical transmission of this type does not occur, but the tsetse flies become diseased and the trypanosomes pass through a part of their life cycle in them. When infected blood is drawn into the fly's stomach, the trypanosomes begin to multiply. After about ten days they migrate forward through the oesophagus into the mouth and through the salivary ducts into the salivary glands. Here they pass through developmental stages that finally take the form of infective trypanosomes. When such a fly bites, some of the trypanosomes pass through the proboscis, and are inoculated into the blood of its victim and thus the infection is transmitted.

The trypanosomes live in the blood plasma of man, not in the blood corpuscles. They multiply rapidly by binary fission and are soon distributed by the blood and lymph to all parts of the body. Trypanosomes seem to be attracted especially by the lymph glands and by the central nervous system and actually penetrate the cells of these organs. This invasion sets up a fever which may reach a height of over 106°F. So many agents are responsible for tropical fever that this symptom is not as important as the enlargement of certain glands, especially those in the posterior triangle of the neck. This is known as Winterbottom's sign, but to the patient its only a "pain in the neck."

1. Human trypanosomes of African sleeping sickness from the salivary glands of the tsetse fly. 2. Swelling of neck glands known as Winterbottom's sign.

FIG. 97. Sanitary officers could prevent the transmission of sleeping-sickness trypanosomes if they could teach tsetse flies not to spit.

In certain cases the patient may be too formidable for the parasites and recovers from the attack, otherwise, if not properly treated, the disease may proceed to the sleeping-sickness stage. The trypanosomes enter the cerebrospinal canal and induce all the symptoms of laziness including a disinclination to exertion, a slow, shuffling gait, a morose, vacant expression, a drooping lower lip and a tendency to lapse into sleep during the day-time without any provocation. Some people might not object to the symptoms up to this point, but continuation of the disease involves many changes too revolting to describe here and the patient finally dies in convulsions or passes into a coma from which he does not awaken. And all this is the result of the bite of a fly that injects a few trypanosomes into the wound with its saliva. The prevention of African sleeping sickness would be simple if we could only teach these tsetse flies not to spit.

HISTORY OF AFRICAN TRYPANOSOMIASIS

Human trypanosomiasis seems to have been recognized first by a British naval surgeon named John (not Tommy) Atkins, who in 1724 published a book entitled "The Navy Surgeon." In the appendix to this book he described "the sleepy distemper which was common among the negroes on the Guinea Coast."

The next mention of the disease occurs in a book written by Winterbottom in 1803 in which the author calls attention to the "species of lethargy which they (the Africans) are much afraid of, as it proves fatal in every instance, . . ." and notes that slave dealers associated enlargement of the glands of the neck with the early stages of the disease, and would not undertake to ship natives who had these swollen glands to America as they knew that they might not survive the voyage and certainly would not live long after it. Thus the name "Winterbottom's sign" became attached to the enlarged glands of the neck.

Trypanosomiasis was redescribed in Sierra Leone by Clark in 1840 who called it "narcotic dropsy." During the next half century a number of papers on the subject were published for the most part by physicians in the French Navy. One of these physicians (Danguise in 1861) called the disease "trypnosis," thereby approximating the present name of the malady, and another physician (Guerin in 1869) discovered that the disease is not contagious.

In the early days many fantastic ideas were advanced to

account for sleeping sickness but none of these were based upon careful experimental evidence. Atkins in his original account states that "the immediate cause of this deadly sleepiness in the slaves is evidently superabundance of Phlegm or Serum, extravased in the Brain which obstructs the Irradiation of the Nerves." After the establishment of the germ theory of disease various bacteriologists attempted to incriminate pneumococci, streptococci, etc., but these pathogenic bacteria simply represented secondary infections.

Other investigators attempted a physiological explanation of the disease describing it as an intoxication due to the eating of various substances. Thus Ziemann as late as 1903 contended that the disease was not an infection, but was due to eating the root of the cassava. Several other investigators (Manson and Mott in 1900) came to the conclusion that human sleeping sickness was in reality a disease resulting from an infection with a parasitic worm, *Acanthocheilonema perstans.*

In the meanwhile, a knowledge of animal trypanosomiases was being built up as noted in Chapter III, which assisted materially in explaining the etiology of human sleeping sickness. Thus before the actual discovery of the trypanosomes of sleeping sickness we knew that trypanosomes existed as parasites in many of the lower animals, that certain diseases of lower animals were caused by trypanosomes, and that the trypanosomes responsible for nagana in horses and cattle were transmitted from animal to animal by tsetse-flies. This knowledge was of great assistance in the elucidation of the relations between trypanosomes and sleeping sickness.

The first trypanosomes reported from human beings were noted by Dr. Forde in 1901. He examined an European who was employed on a government steamer on the Gambian river and was suffering from what was considered to be malaria fever. No malaria parasites were found in his blood but worm-like organisms were seen whose nature was unknown to Forde. Later in the same year Drs. Dutton and Forde together examined the patient and recognized the worm-like organisms as trypanosomes. Dutton then gave an excellent description of the protozoon and named it *Trypanosoma gambiense*. These observations demonstrated the presence of trypanosomes in man, but did not prove their pathogenicity.

In 1902, Dutton and Todd undertook further studies on human trypanosomiasis in the Senegambia. Of 1000 individuals examined six natives and one half breed were found to be infected, but since they appeared to be in almost perfect health they concluded that the infection in the African native produced a particularly mild "trypanosome fever." They did not associate trypanosome fever however with the fatal sleeping sickness.

An epidemic of sleeping sickness that spread rapidly in Uganda at this time took toll of thousands of lives as it gradually crept up the newly established trade routes of the Congo basin. Something had to be done about this so in 1902 a Sleeping Sickness Commission, comprised of Drs. Low, Christy, and Castellani was sent to Entebbe, the capital of Uganda. Drs. Low and Christy following a suggestion of Sir Patrick Manson, tested without results the hypothesis that sleeping sickness was due to a blood-inhabiting worm. Dr. Castellani examined the cerebro-spinal fluid of many patients and reached the conclusion that a streptococcus was the cause of the disease. He saw trypanosomes occasionally but believed from the work of Dutton and Todd that they had nothing to do with the sleeping sickness.

Drs. Low and Christy had gone back to England and Castellani was on his way when, in 1903, Sir David Bruce was sent to Uganda. Castellani told Sir David about the streptococcus and also about the trypanosomes present in the cerebral spinal fluid of patients dying of the disease. Bruce having studied trypanosome diseases in horses and cattle realized the significance of this discovery and persuaded Castellani to return to Entebbe with him. Together they examined about forty patients already semi-comatose with sleeping sickness and found trypanosomes in all of them. Castellani who returned to England three weeks later announced that he had discovered a trypanosome to be the etiological agent of African Sleeping Sickness. The credit for this discovery, consequently, is usually, though erroneously, attributed solely to Castellani, although it was due mainly to Bruce.

Sir David and Lady Bruce and their coworkers subsequently established the fact that trypanosome fever is the first stage of human trypanosomiasis and is due to the same trypanosome as that of sleeping sickness and that certain species of tsetse flies of the genus *Glossina*, mainly *G. palpalis*, transmitted the trypanosomes from man to man. The life cycle of the trypanosome in the tsetse fly was worked out by Kleine in 1908.

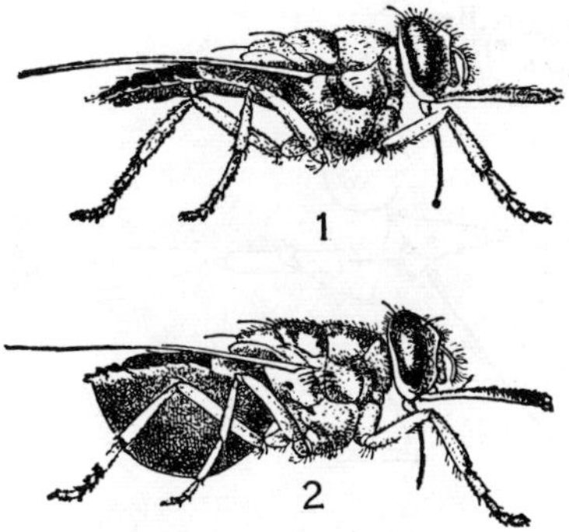

FIG. 98. Tsetse flies (*Glossina palpalis*), the transmitting agent of African sleeping sickness. (1) Before feeding; (2) after feeding.

ATTEMPTS TO CONTROL SLEEPING SICKNESS

The discovery of the trypanosomes of sleeping sickness and of their method of transmission furnished all of the ammunition necessary to declare war on the disease. The first battle, which was fought on the shores of Lake Victoria-Nyanza, was a strategic retreat. Tsetse flies were known to live only in the lowlands surrounding the lake hence transmission could be avoided by transferring the natives to some other region. Soon, however, sleeping sickness was reported from Nyassaland and Rhodesia where the *palpalis* type of tsetse fly was not known to live.

The life of a scientist is not always a happy one since his work, like that of the old-fashioned housewife, never seems to be done. Accordingly an investigation had now to be made of the Rhodesian type of sleeping sickness. The disease proved to be more acute and fulminating than the Uganda variety and the trypanosome responsible for it was by some protozoologists believed to be a separate species although it seems to be almost as closely related to the other species as a Siamese twin. Another species of tsetse fly, *Glossina morsitans,* was discovered to be the transmitting host. No new

FIG. 99. What we need to eliminate sleeping sickness is to destroy the tsetse flies that transmit it—who cares how this is accomplished?

types of sleeping sickness have been reported since 1909, hence public health activities have been concentrated on the attack of the Gambian and Rhodesian varieties.

Two principal methods of preventing the spread of sleeping sickness are recognized; one is to attack the fly and the other to make it impossible for the fly to become infected. Unfortunately, man shares his susceptibility to sleeping-sickness trypanosomes with certain lower animals, especially antelopes. The situtunga antelope is commonly infected in nature but exhibits no symptoms nor injury. Oxen, goats and sheep may also serve as reservoirs from which tsetse flies may acquire infections. Such a situation renders it practically impossible to prevent flies from becoming diseased.

Attempts to eradicate the tsetse flies were begun as soon as they were recognized as transmitting hosts. This has been accomplished in certain limited localities, but on the whole the flies have successfully protected their goal line. It is possible to prevent infection in a group of natives by moving them *en masse* to a region known to be free from flies, but one can hardly consider this much of a victory, especially since the flies remain infected and are able to transmit the trypanosomes to anyone who may enter the evacuated territory. Recently an ingenious but dubious method of exterminating the flies has been suggested. A dummy bull is set up and lighted within so that the light shines through a hole in its side. Flies that approach in order to bite the bull are attracted by the light, enter through the hole, and presumably are unable to find their way out.

TREATMENT OF SLEEPING SICKNESS

Greater success has attended efforts to control the symptoms of sleeping sickness and to cure the disease after infection has become an accomplished fact. The two drugs that at present appear to give the best results are Bayer 205 and Tryparsamide. Bayer 205 was elaborated by German chemists, and, after experiments on trypanosomiases in lower animals, was introduced in 1920 as a remedy for sleeping sickness. The effects of this drug are remarkable. The trypanosomes are usually no longer visible in the blood after twelve hours and the physical and mental condition of the patient undergoes a marvelous improvement after the first dose. Cures have been reported in about 80 per cent of cases, but when once the trypanosomes have invaded the gray matter of the brain they cannot be destroyed by this drug.

An intriguing story is told of how the Germans offered to exchange the formula of Bayer 205 to France in return for some of the territory in Africa that had belonged to Germany before the World War, and which had been given to France after the war. The land was fertile, but almost useless, because of the presence of sleeping sickness. Instead of making the trade, the French went to work in their own laboratories in an attempt to discover either the composition of Bayer 205 or to develop a drug that was just as good. Whether or not this story is true, the French produced a drug called Fourneau 309, that is equivalent to Bayer 205 and has met with similar success.

The other drug that has been proved to be most effective against sleeping sickness is known as tryparsamide. This compound was elaborated by Brown and Pearce at the Rockefeller Institute. The French also have an equivalent for this drug which they call "Fourneau 270." Tryparsamide is particularly valuable in cases in which the nervous system is involved since it apparently may penetrate the substance of the brain and destroy the trypanosome invaders there as well as in the cerebro-spinal fluid.

One difficulty encountered by those who have attempted to cure sleeping sickness is the tendency of the trypanosomes to become drug-fast, that is, certain of the trypanosomes that are not killed at once become resistant to the drug. In cases of this kind one drug is administered for a time and then the other is substituted.

CHAGAS DISEASE IN AMERICA

African sleeping sickness is of greater importance as a world problem than Chagas disease but the latter is of more direct concern to us because it occurs in America and with the perfection of rapid transportation between the American Republics we may expect the spread of American trypanosomiasis at least to the southern part of this country. The transmitting agents of Chagas disease are insects of the kissing bug type that belong to the genus *Triatoma*. Species belonging to this genus live in the southern States, as far north as southern Illinois and Ohio, and also in the southwestern States. A species of kissing bug (*Triatoma protracta*) that lives in the nests of wild rats in California is infected in nature with trypanosomes indistinguishable from those of Chagas disease. Trypanosomes from these bugs, when inoculated into opossums, wood rats and albino laboratory rats, brought about successful infections. If you come from California and have tears, prepare to shed them now because the stage is all set for the spread of Chagas disease to human beings in California, provided the inhabitants there are not already infected, since

FIG. 100. Not a moving picture, but a scene near Hollywood where susceptible human beings, rats, bugs, and trypanosomes have set the stage for a nice epidemic of Chagas disease.

the trypanosomes, the bugs that transmit the organisms, the reservoir hosts (opossums and wood rats), and susceptible human beings are all living there together under favorable conditions. (Florida newspapers please copy.)

The trypanosome of Chagas disease resembles other trypanosomes in general appearance but has a very large parabasal body near the posterior end. Its life cycle in man differs in several respects from that of the sleeping-sickness trypanosomes. In the first place multiplication does not take place in the blood stream, and in the second place the trypanosomes invade certain organs, such as the heart, muscles, thyroid gland, lungs, etc., where they become oval in shape. In this form they are known as leishmania bodies. These bodies undergo binary fission and in course of time cysts containing hundreds of them develop. The leishmania bodies may transform into trypanosomes and break out into the blood stream; these in turn may invade the tissues again and become leishmanias.

Trypanosomes of Chagas disease as they appear in the blood.

When a bug bites an infected person, the trypanosomes are drawn with the blood into the stomach where they multiply by binary division. After certain changes take place in them,

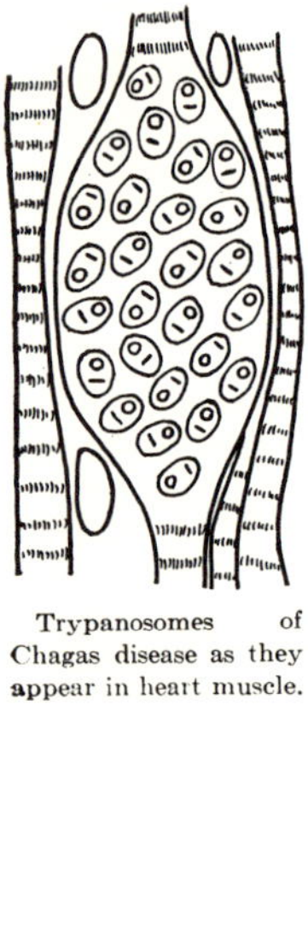

Trypanosomes of Chagas disease as they appear in heart muscle.

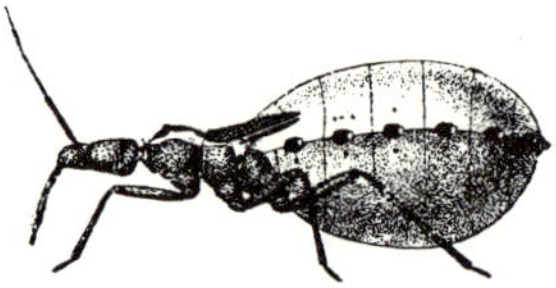

FIG. 101. Kissing bug (*Triatoma megista*), the transmitting agent of Chagas disease.

they pass on into the rectum of the bug and out of the body in its excrement. The bugs are active at night usually biting a person on the neck. During the process of biting and sucking up blood the bug may evacuate a drop of excrement containing trypanosomes. The bite is irritating and the person bitten naturally rubs the spot, thereby forcing into the wound some of the trypanosomes. Infection of susceptible human beings results by this so-called contaminative method.

The effects of the infection depend on the organs invaded by the parasite; thus when the heart is parasitized a myocarditis results, and when the nerve cells are invaded nervous symptoms arise. Irregular fever, enlarged glands and anemia are usually the general symptoms.

As in the African trypanosomes, reservoir hosts are largely responsible for maintaining the race, since opossums and armadilloes are susceptible and may furnish an infective dose to any bug that chances to bite them.

No satisfactory treatment for Chagas disease has been discovered. Drugs such as Bayer 205 and tryparsamide have no effect upon it. Fortunately the disease is usually mild.

THE HISTORY OF CHAGAS DISEASE

The history of Chagas disease is of considerable interest since the etiological agent was discovered in the invertebrate host before the disease it produced was known to occur in man. While carrying on an anti-malarial campaign in the interior of Brazil, Dr. Carlos Chagas noted large bugs in the huts of the natives. The digestive tract of these blood-sucking insects was found to be heavily infected with trypanosomes. Some of the bugs were sent to Dr. Oswaldo Cruz who allowed them to feed on a monkey in whose blood trypanosomes were found later. Suspecting that the same trypanosome might be present in the natives of the region, Chagas examined the blood of a large number of the inhabitants and finally found them in a sick child whose symptoms were subsequently proved to be due to the trypanosomes.

HUMAN LEISHMANIASIS

Trypanosomiasis and leishmaniasis are nice long terms whose pronunciation gives one a feeling of singular satisfaction. They are usually considered together because the stages in their life cycles indicate a near relationship. For example, the organisms responsible for leishmaniasis are indistinguishable from the leishmania stage in the life cycle of the trypanosome of South American trypanosomiasis. The leishmanias like the trypanosomes are rather spotty in their distribution. Some of them occur in the Near and Far East and others in America reminding us that "East is East and West is West." Three or four types of disease due to these organisms are usually recognized. Kala-azar is perhaps the most important. The inhabitants of large areas in China, India, other parts of Asia, Africa and southern Europe suffer from this affliction. Oriental sore is somewhat similar in geographical distribution. American leishmaniasis, or espundia, occurs principally in South America, but is known to extend northward through Central America into Mexico and may at any time be expected to appear in the United States.

HISTORY OF HUMAN LEISHMANIASIS

The history of the discovery of the leishmanias does not cover as long a period as that of the discovery of the trypanosomes, but emphasizes a feature of scientific research that is exceedingly interesting and not uncommon, namely, the simultaneous investigation of an organism or a disease, that had never before attracted particular attention, by several scientists working in different localities and without knowledge of each other's activities. In the case of leishmaniosis, the years 1903 and 1904 were particularly noteworthy. Cunningham in 1885 had reported on "Peculiar parasitic organisms in the tissue of a specimen of Dehli boil" and actually saw leishmania bodies in white blood corpuscles or macrophages but he thought he was observing amebas containing spores. This observation was confirmed by Firth in 1891 who made the same mistake. He gave these macrophages the name *Sporozoa foruncculosa.*

Nothing of note happened between this date and 1903. Then on May 30 of that year Leishman published a paper on "the possibility of the occurrence of trypanosomiasis in India" in which he described rounded bodies of a parasitic nature in patients suffering from kala-azar or dum-dum fever. Donovan on July 11, 1903 recorded independently the same bodies in the same disease. Donovan's slides were examined by Laveran and Mesnil at the Pasteur Institute in Paris and they decided that these bodies belonged to the same genus as the parasites of Texas fever in cattle; on November 3, 1903 they proposed the name *Piroplasma donovani* for them. Eleven days later, on November 14, 1903, appeared a paper by Sir Ronald Ross in which he expressed the belief that they were Sporozoa and proposed for them the genus name *Leishmania.*

Wright, working in Boston, reported in December 1903 the presence of similar bodies in a cutaneous ulcer on the body of an Armenian child which he located tentatively in the order Microsporidia and to which he gave the name *Helcosoma tropicum.*

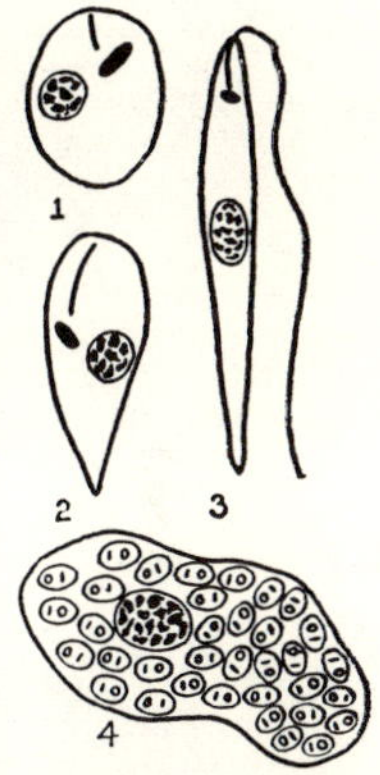

Leishmanias. 1 and 2. Parasites of oriental sore from a sore. 3. Same from a culture. 4. White blood corpuscle filled with parasites of kala-azar.

The year 1904 was almost as productive as 1903 in the scientific progress of leishmaniasis. Marginowsky and Bogroff noted bodies similar to those described by Wright in a sore on the body of a resident of Persia. Bentley recorded what were by that time known as leishman-donovan bodies in cases of kala-azar, and Sir Leonard Rogers succeeded in cultivating these bodies in vitro in this way demonstrated their flagellate nature. Leishman-donovan bodies and trypanosomes were thus shown to be closely related. According to the rules of prority regarding scientific names the protozoan parasites of dum-dum fever, or kala-azar, are known as *Leishmania donovani*, and those of oriental sore, as *Leishmania tropicum.* The organisms of the cutaneous leishmaniasis, or espundia, prevalent in South and Central America were described by Vianna in 1911 and named *Leishmania braziliensis.*

KALA-AZAR

More problems remain unsolved in leishmaniasis than in most other protozoan diseases, largely due to the fact that the method of transmission is not definitely known. Perhaps the best method of bringing these to your attention is to describe the organisms and then discuss our knowledge of each of the leishmania diseases separately.

The organisms responsible for the leishmaniases appear in man only in the leishmania form, that is, as minute oval structures containing a nucleus and a parabasal body. When these leishmanias are placed in a proper culture medium they transform into slender flagellates resembling trypanosomes but lacking an undulating membrane. The flagellate form also occurs in certain insects. Kala-azar is sometimes called visceral leishmaniasis because the internal organs of the host are invaded, in contrast to oriental sore, or dermal leishmaniasis, in which only the skin is parsitized.

No one knows with certainty how kala-azar is transmitted from one human being to another or whether the invading organisms are in the flagellate or in the leishmania stage. We do know that the phagocytic cells, which ordinarily protect the body from such invading parasites, are largely responsible for the development of the disease. The parasites are engulfed by phagocytes that line the capillaries, but the phagocytes seem unable to destroy them; on the contrary, the leishmanias grow and multiply rapidly at the expense of the phagocytes and finally completely fill these cells. Then the phagocyte bursts and a horde of leishmanias are liberated into the blood stream. The most conspicuous symptoms of kala-azar are enlargement

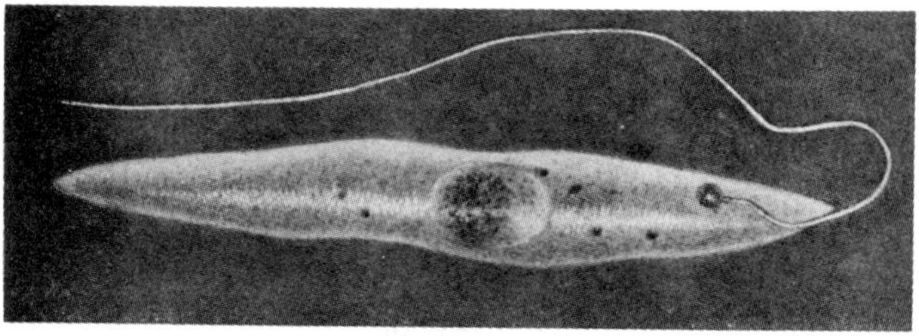

FIG. 102. The flagellate stage in the life cycle of the leishmania parasite of kala-azar and oriental sore as it occurs in test tube cultures.

of the liver and spleen, fever, dysentery, emaciation, and a change in the skin to an earthy-gray color. The native name for kala-azar disease or "the black disease," is based on this change in color.

Since about 90 per cent of kala-azar patients succumb to the disease if not treated, an effective drug is very desirable. A drug that was reported by Vianna in Brazil in 1912 to be specific for dermal leishmaniasis was soon found to be equally valuable in the treatment of kala-azar. This was antimony tartrate, or tartar emetic. Many compounds of antimony have since been employed with great success so that a 90 per cent mortality has been changed to an equally high rate of recovery.

Oh, East Is East and West Is West

ORIENTAL SORE

The dermal leishmaniasis of the Old World is known generally as oriental sore, although many local names are applied to it such as Dehli boil by the inhabitants of Bagdad; the natives of Dehli return the compliment by calling it Bagdad sore. The leishmanias of oriental sore are indistinguishable from those of kala-azar but are limited to small areas of the skin in which sores develop. Within a few months healing takes place

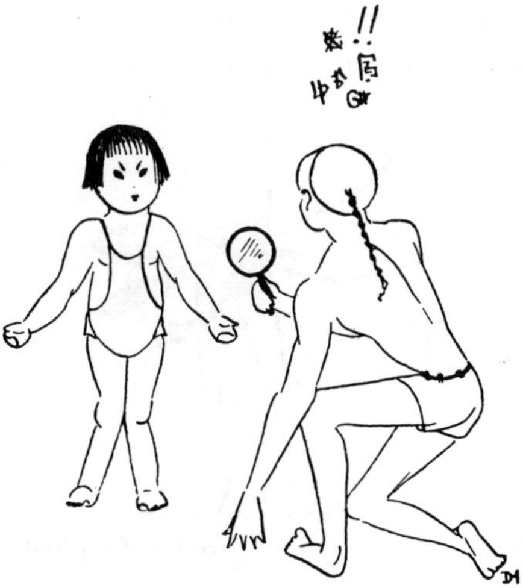

FIG. 103. "Where," says this distracted mother, "can I find a spot to vaccinate against oriental sore that will not be exposed when my little girl grows up?"

but an ugly scar is left behind. One of the unusual features of this disease is the very definite immunity that results from infection; that is, a person who has once been infected is no longer susceptible to the disease. On this account certain orientals have adopted the custom of infecting their children artificially on some part of the body that is ordinarily unexposed so as to avoid disfiguring scars on the face. Unfortunately modern styles of dress are making it more and more difficult to find an area of the body that is not exposed.

ESPUNDIA

South American leishmaniasis is no such harmless disease as oriental sore. To the natives, who call it espundia, uta or forest yaws, it is a terrible scourge. The disease is particularly prevalent among men who work in the forests, especially in the chichleros who go out in search of chicle which is used in the manufacture of chewing gum. This leishmania has a particular preference for the mucous membrane of the mouth, nose and throat. Here it brings about a chronic ulceration of the tissues and the destruction of the bones and cartilage of the nose and mouth. Proper treatment with tartar emetic

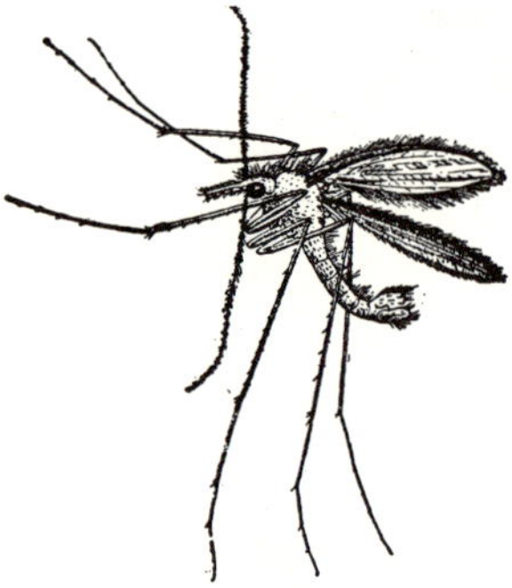

FIG. 104. Sandfly, supposed vector of oriental sore

or with one of the antimony compounds relieves the patient of the most distressing symptoms and eventually cures the disease.

TRANSMISSION OF LEISHMANIASIS

The most exasperating aspect of the leishmania problem is that of transmission. The amount of time and effort spent on this subject by protozoologists and physicians of high standing during the past thirty years is difficult to estimate. Because certain other diseases, such as malaria and sleeping sickness, are transmitted by blood-sucking insects, investigators have naturally attempted to incriminate insects that infest man in regions where the leishmaniases occur. The private life of the bedbug has probably been probed more thoroughly than that of any other suspect. According to an old song "the bedbug hasn't any wings at all but he gets there just the same." This is good natural history since although bedbugs are wingless they are present wherever there are human beings. However, in spite of his evil reputation, the bedbug appears to be as innocent of wrong doing as a babe unborn.

On the other hand, leishmanias have been found to live readily in sandflies, but if, as and when they are conveyed to the vertebrate host is still as uncertain as whether the stock market will go up or down. However, its an ill wind that blows nobody good, and as long as this predicament continues there will be plenty of work to keep protozoologists and entomologists out of the bread line.

KALA-AZAR IN AMERICA

One of the most interesting discoveries regarding American leishmaniasis has recently been announced. In 1932 an instrument named the viscerotome was devised in Brazil for the purpose of obtaining a small sample of liver tissue from human bodies on which an autopsy could not be performed, in order to determine whether or not death was due to yellow fever. Microscopic study of these samples has revealed eighty-five cases of visceral leishmaniasis indistinguishable from kala-azar. Clinically patients suffer symptoms similar to those of kala-azar in Asia but no changes in skin-color occur. The disease appears to be scattered widely in Brazil and in the Argentine Chaco and probably would be found to exist further north in Central America and possibly in the southern United States if proper examinations were made. Many cases of fever accompanied by enlargement of the liver and spleen are encountered in the American tropics that are not successfully diagnosed. Perhaps some of these are due to visceral leishmaniasis. Thus our knowledge of this group of protozoan diseases continues to increase from year to year.

INTRODUCTION OF TROPICAL DISEASES INTO THE UNITED STATES

Before concluding this chapter I wish to stress two points; first, the important relation of these diseases to colonization in the tropics, and second, the danger of introducing them into the United States.

The dream of an International Highway from the frozen Arctic to the southern tip of South America is fast developing into reality. During the past three summers I have traveled by motor car from Guatemala to Honduras and El Salvador,

FIG. 105. Tourists on the Pan-American highway will probably transport Chagas disease and other tropical diseases to the United States.

from Guatemala to close to the Mexican border and from Panama to Costa Rica. Steady progress is being made and it will not be many years before we shall be able to visit the home of tropical diseases without the danger of becoming seasick. What is even more interesting is the possibility, and even probability, that it will not be necessary for students of protozoology to travel to Central and South America and the West Indies for their material, since protozoa will be brought to them.

Travel by motor bus in the republics of Tropical America is easy and inexpensive. A large proportion of the population can afford to ride from one country to another, and the motor buses are usually so crowded with a heterogeneous mob of natives that Americans and other foreigners hire private cars. No quarantine regulations are now in force against any tropical disease due to pathogenic protozoa, and such regulations in the future will be practically impossible. It seems inconceivable that adequate examinations for the infective stages of organisms, such as those responsible for Chagas disease, espundia, malaria and amoebic dysentery, will ever be made at any international border. We may expect, therefore, to entertain in the United States patients with diseases that at present are hardly more than names to us. We may also expect a certain amount of transfer of infective organisms from these foreigners to nonimmune Americans.

Travel in the United States from north to south furnishes a somewhat similar problem. At present, almost everyone who has sufficient time and can afford to do so, leaves home when summer arrives and motors about the country. When the International Highway is completed, the lure of the Tropics will tempt a considerable proportion of these tourists to Mexico, Guatemala, Panama, Venezuela, Brazil and other Tropical American countries. Many of them will remain there long enough to become infected with various protozoa and will be just about ready to exhibit symptoms when they reach home. This means that sporadic cases of tropical diseases will appear in various parts of the United States where they had never been encountered before; and that, where conditions are particularly favorable, epidemics may be expected. It is even easy to imagine new endemic areas being established in localities where sanitation is of a low order or satisfactory intermediate hosts are indigenous.

Everyone knows how often under conditions as they exist at present, tropical diseases that occur infrequently in temperate regions are diagnosed incorrectly. A number of my friends have been subjected to consider inconvenience and have been placed even in danger of their lives because of the failure of their physicians to recognize such well-known diseases as malaria and amebic dysentery. The early and correct diagnosis of these and other tropical diseases will be much more important in the future than it is now; and steps will have to be taken to prepare for the changed conditions that are almost certain to arrive within a few years.

Interest in tropical medicine has been increasing rapidly in this country in recent years. The American Society of Tropical Medicine is a flourishing organization. The American Society of Parasitologists and the American Microscopical Society are both devoted largely to the sciences underlying tropical medicine. Recently the American Academy of Tropical Medicine was organized with the object of coördinating and stimulating work in this field. These activities seem to me to be fully justified and should have far-reaching results.

If I were to summarize this chapter in the form of a poem, it would bear the title "Oh! East is East and West is West;" if you read this during the Christmas season I earnestly urge you to sing it.

We two bugs of orient are,
Spreading woe both near and far.
Sores on neck and face deliver,
Or we may enlarge your liver.
Sandflies, pesky little vectors,
Squirt you full with their injectors.
Leishman his good name did render
To the iases we engender.
South America too has bugs,
Cousins of the Leishman thugs.

Other germs in Afric flourish,
Sleeping sickness they encourage.
Tsetse flies transmit the villians
From antelope to native millions.
Trypanosomes they call each other;
Their iases few recover.
Chagas found another villo
In kissing bug and armadillo;
Traced it to a human sore;
Wrote his name forever more.

Chapter IX

WINGED GUINEA PIGS

IF YOU believed in reincarnation and wanted to be certain of plenty of adventures after your next transmigration you should pray to be born again as a malaria parasite. Here is a living speck of protoplasm smaller than a red blood corpuscle that in the short period of one month may circulate gaily through the body of a human being, then be transported hither and thither on the wings of a mosquito, and, in course of time, be delivered once more into the lifeblood of man. During this period you would become intimately familiar with human anatomy, and would learn how mosquitoes make whoopee, but you would be extremely lucky to arrive again in the bosom of *Homo sapiens,* for this adventure would indeed be perilous and suitable only for those willing to undertake a hazardous enterprise. Nature is more prodigal of malaria germs than a bride is of blushes; when a human patient dies 6,000,000,000,000 malaria parasites die with him, that is, 100 times as many parasites as there have been seconds since the birth of Christ; and only one out of millions of mosquitoes ever finds an infected person to bite or succeeds in transmitting the infection to another susceptible human being.

LIFE CYCLE OF HUMAN MALARIA PARASITE

Malaria parasites can be conveyed from one person to another in the laboratory by blood inoculation, but in nature female mosquitoes belonging to a certain species of the anopheline tribe have cornered the market and have secured a monopoly on this nefarious racket. Yes indeed "the female is more deadly than the male." "Truth is stranger than fiction" in the animal world as well as among us mortals. Certainly no one could possibly have imagined such a strange life cycle as that of the malaria organism in man and mosquito. The stages in this cycle are illustrated in the accompanying diagram. Here in outline is shown a section of a human blood vessel and the business end of an anopheline mosquito. With this before us we may pass in review the marvelous series of events that crowd swiftly on one another as the drama of this life cycle unfolds. This has always seemed to me to be one of

FIG. 106. That famous mythological parasitologist, Jet Wimp, dreams of the adventures he might have if he were reincarnated as a malarial parasite.

the most remarkable adventures in living experienced by any species of animal. Such a life cycle nourishes the sort of respect for nature expressed by the theological student who remarked that when he looked through the microscope he could see the finger prints of God on every atom.

The infective stage in the mosquito is known as a sporozoite, a minute spindle-shaped cell ten one-thousandths of a millimeter in length. The sporozoites are stored in the salivary glands of the insect. When the mosquito "bites" it pierces the skin with its proboscis and a salivary secretion containing the parasites then passes along a groove and into the wound. The red blood corpuscles seem to be entered by boring movements of the sporozoite—a process that requires from about forty to sixty minutes. Probably not all of the sporozoites that are

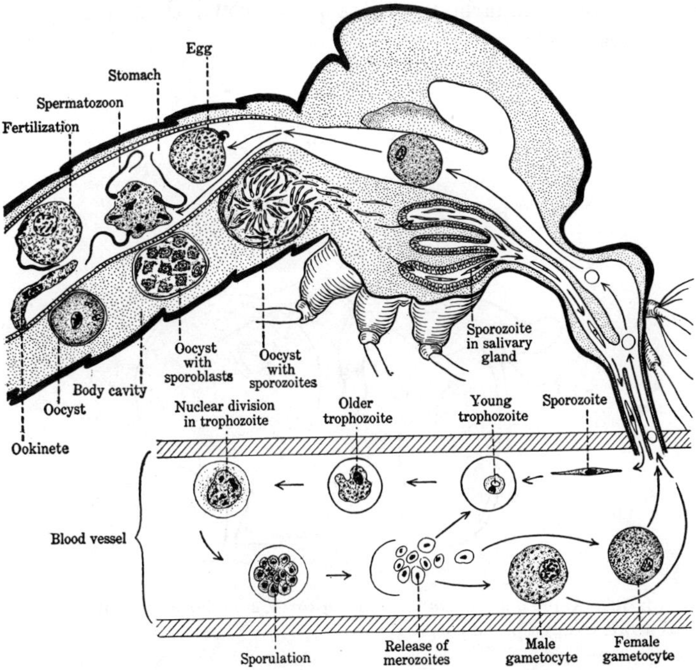

FIG. 107. By following the arrows in this diagram one can trace the life cycle of a malarial parasite in a human blood vessel and in the body of a mosquito. A good place to start is with the sporozoite where the proboscis of the mosquito enters the blood vessel.

inoculated by the mosquito succeed in stealing a ride in a red blood corpuscle, since many are no doubt engulfed by those ever-vigilant traffic cops of the blood-stream, the phagocytes.

The number of sporozoites injected into a human being by a mosquito depends on the number present in the salivary glands of the mosquito and the number of times the mosquito bites. There are perhaps not more than 10,000 sporozoites in the salivary glands of a single mosquito and usually less than this number. That several thousand may enter the blood from a single mosquito seems quite reasonable. Sometimes the blood cells are multiple-infected,—that is, two, three, four or more malaria parasites may crowd their way into a single corpuscle.

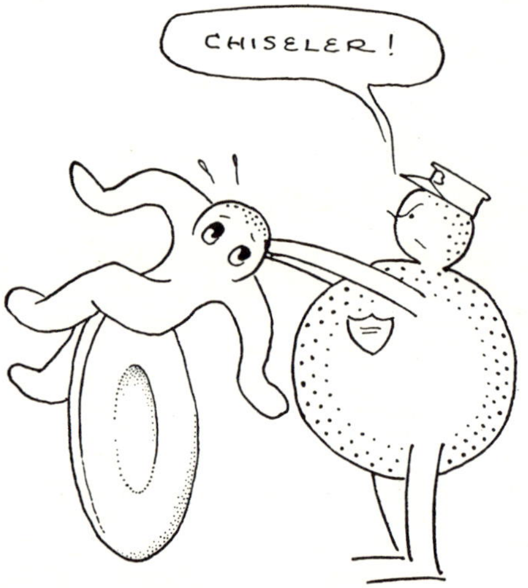

FIG. 108. A phagocyte traffic cop captures a malaria parasite trying to steal a ride in a red blood corpuscle.

The sporozoite changes into an ameboid form after it enters the blood corpuscle and is then known as a trophozoite. When living blood is examined under a microscope the trophozoite may be seen actively throwing out and withdrawing pseudopodia within the corpuscle. The ungrateful parasite not only uses the blood corpuscles as a means of transportation but also treats it as a meal ticket, gradually eating the highly nutritious contents. As the growth of the parasite proceeds dots appear in the protoplasm of the red corpuscle. After six or eight hours of continuous feeding, fine, reddish-brown granules begin to appear within the greedy parasite. These are pigment granules (also called melanin or haemozoin), which probably represent those constituents of the hemoglobin that the parasite is unable to digest. The young trophozoite when stained appears ring-shaped. Older trophozoites change from the ring stage to an ameboid shape and by the end of forty

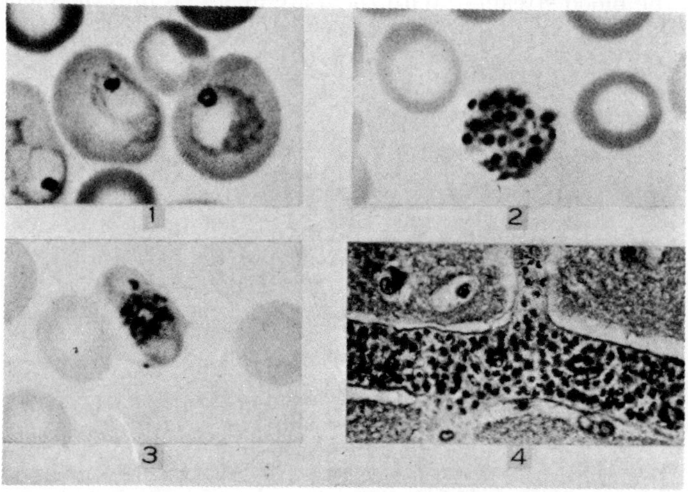

FIG. 109. Stages in the life cycle of human malarial parasites are shown in these microphotographs. (1). These trophozoites occupy almost half of the three red blood corpuscles they are living in. (2). A malarial parasite is segmenting into about twenty merozoites. (3). A crescent-shaped sexual stage. (4). Sometimes the capillaries in the brain become occluded by malaria parasites.

hours the parasites have practically replaced the protoplasm in the red blood corpuscles by what they no doubt believe to be a superior type of living substance,—a sort of Nordic protoplasm.

At this time the demands of the race overcome those of the individual and reproduction occurs. Trophozoites when they are ready to sacrifice themselves for the good of the race are

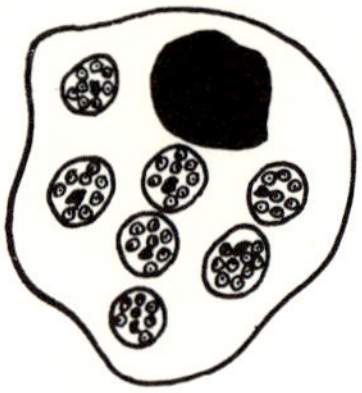

A phagocyte that has engulfed seven parasitized red blood corpuscles.

called schizonts. The fully-grown schizont at the end of forty hours undergoes several successive divisions and gives rise usually to from eighteen to twenty minute cells called merozoites. This process of schizogony, or "segmentation," requires about six or eight hours. The merozoites are active youngsters that don't like to be caged up at home so they break out of the red blood corpuscles and go wandering about in the blood stream. With them are liberated into the blood,

FIG. 110. The phagocytes in the blood stream guard the human host against such wicked foreigners as malaria parasites.

pigment granules, the remains of the blood corpuscles, and probably poisonous substances produced by the parasites.

The merozoites don't lead a nomadic life for very long but soon become attached to and penetrate fresh red corpuscles. In some unknown manner they are attracted to young corpuscles and seldom enter adult cells although only about one per cent of the red corpuscles are of the young type. After they enter the young corpuscles they become trophozoites and proceed to repeat the asexual cycle. Many of the merozoites are probably destroyed by phagocytes, or otherwise, during their period of vagrancy and probably only about half of them succeed in growing up and reproducing. This is a much higher death rate than is known to occur among human beings anywhere. Evidently malaria parasites might thrive even better than they do if they would train their young to avoid those big, bad phagocytes.

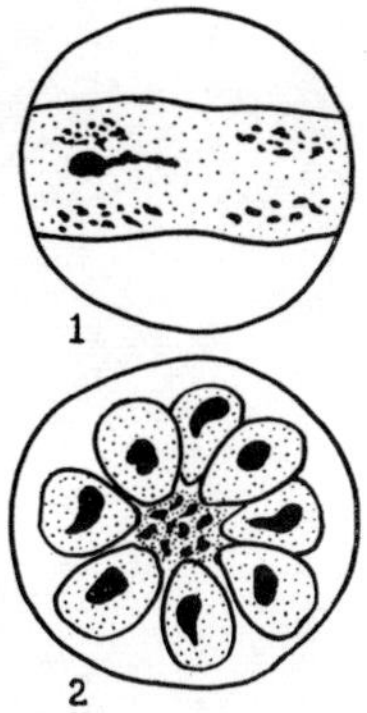

Parasites of quartan malaria in red blood corpuscles. 1. Trophozoite. 2. Segmenter.

The interval between the inoculation of parasites by the mosquito and the appearance of chills and fever is called the period of incubation. Chills and fever are due to the breaking down of the red blood corpuscles but do not appear until a sufficiently large number of parasitized corpuscles break down at one time—a number estimated by Ross as 150,000,000 in an average man. Ordinarily the period of incubation in benign tertian malaria is about two weeks.

After the infection has continued for a few days, a new stage of the parasite begins to appear in the blood. This is the gametocyte or sexual stage. Two kinds of gametocytes are formed, female, or macrogametocytes, and male, or microgametocytes. These are all the stages in the life cycle that occur in the blood of human beings.

The parasites I have been describing are those of benign tertian malaria. The parasites of quartan malaria are smaller and often bandlike and produce fewer merozoites; those of malignant tertian malaria form very small rings and often one red cell is infected by two or more parasites. The gametocytes of malignant tertian malaria are shaped like a crescent.

Parasites of opposite sex take no interest in each other as long as they live in the human host, and probably disintegrate in the blood when they are a week or two old. They sit up and take notice, however, if they are sucked into the stomach of a mosquito. Then the microgametocyte immediately gets all upset, and, after a period of intense internal activity, from six to eight flagella-like processes burst out; these are microgametes and their formation is known as "exflagellation."

The microgametes are obviously very much agitated by their sudden freedom and lash about violently until they break away from the mother cell and swim about wildly in the blood.

In the meantime the macrogametocyte is laboring industriously to fulfill her mission in the world, which is accomplished when her body has become transformed into a single large female sex cell, the macrogamete. As in most higher animals the macrogamete is large and passive and the microgamete small and active. Since the microgametocyte produces six or eight microgametes, or male sex cells, and the

FIG. 111. Less than one out of every six male parasites (microgametes) ever succeed in mating with a female (macrogamete), all of the others are destined to be bachelors.

macrogametocyte produces only one macrogamete, or female sex cell, there are on the average about six times as many male cells as there are female cells. Furthermore, since only one male cell mates with a single female cell, most of the male cells are destined to remain batchelors as long as they live. They do their best, however, to find a mate as they wriggle about energetically among the blood corpuscles. In course of time each macrogamete acquires a microgametic husband, who is completely engulfed by his mate. The decision at the end of this unequal contest must be given as usual to the female of the species whose cannibalistic achievement transforms her into a fertilized egg. The egg is at first spherical and passive but soon becomes elongated and active and is now called an oökinete, or sometimes a vermicule. The oökinete moves about like a worm finally boring its way into the stomach wall

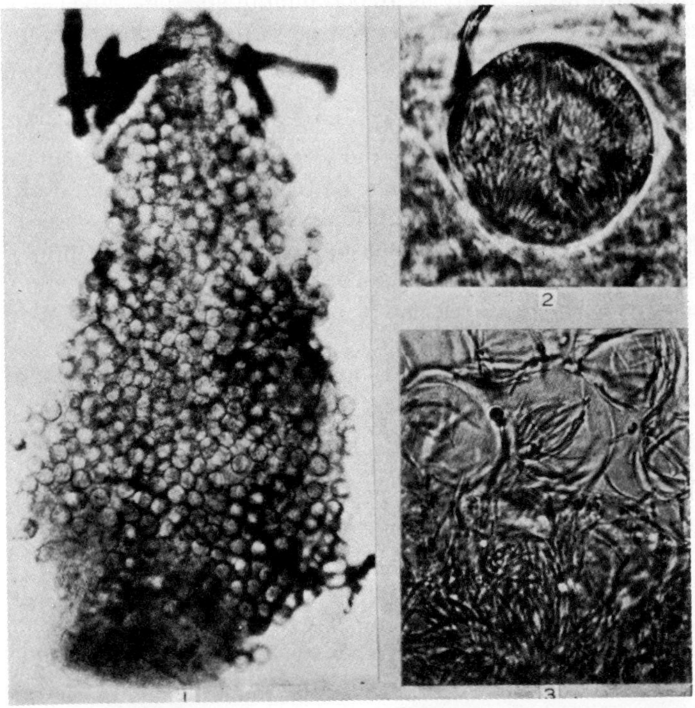

FIG. 112. (1) The outer surface of the stomach of a mosquito heavily infected with malaria parasites is covered with hundreds of oöcysts. (2) A single oöcyst very highly enlarged reveals the sporozoites within. (3) Slender spindle-shaped sporozoites in the salivary glands of a mosquito.

of the mosquito where it once more becomes a passive spherical body, known now as an oöcyst.

The process of fertilization, the change to the oökinete, and the formation of an oöcyst occupy about forty hours at ordinary summer temperature. The oöcyst, which lives a life of ease at the expense of the surrounding cells, grows enormously in size and eventually projects out like a knob, not into the stomach, but into the body cavity of the mosquito. Inside of the oöcyst the nucleus divides repeatedly and the protoplasm finally forms as many as 10,000 minute spindle-shaped sporozoites. The oöcyst, having now performed its function, bursts and the sporozoites escape into the body cavity of the mosquito. This cavity is not like the body cavity of man but is a hemocoel filled with blood. The blood circulates through it and bathes the various internal organs, and in this circulating blood the sporozoites are carried to all parts of the mosquito's body. They do not remain long in the blood but proceed to bore their way into various organs, but especially into the salivary glands. Here they wait in a sort of cold storage always ready to be injected into any luckless human being the infected mosquito may chance to bite.

We have already noted how extraordinarily rapidly the numbers of the fresh-water ciliate, paramecium, increase by simple binary division; but paramecium may be considered practically sterile when compared with malaria parasites. The latter, however, must pay for the joys of an adventurous career by excessive mortality, since only one malaria parasite in millions can expect to be sucked up by a mosquito or injected into a susceptible human being, hence reproductive powers must be stepped up to an almost unbelievable degree or the malaria race would soon disappear so completely we couldn't even make a national monument out of it.

BIRD MALARIA

Baltimore was once an excellent city in which to study malaria and several important discoveries were made and a number of large monographs were published in the gay nineties by students and members of the staff of the Johns Hopkins Medical School. But now a patient suffering from malaria is a *rara avis* and investigations on human malaria have been ruined by the disappearance of the infection. If we wish to carry on research in this field, we must therefore, migrate to some malarious country or use lower animals that are susceptible to malaria. As a matter of fact we do both, but serious laboratory work is now usually done with bird and monkey malaria not only at Hopkins but also in most of the other institutions both in the United States and abroad where malaria is studied.

Many species of birds, even those that "sing in the spring tra la," suffer from malaria, but the organisms with which they are infected are specifically different from those that live in man. No one has ever been able to infect birds, even under the most favorable laboratory conditions, with human malaria

FIG. 113. Culex mosquitoes squirt us full of bird malaria parasites

germs; and man and other mammals seem equally resistant to the parasites of birds. The organisms of bird malaria are transmitted by culicine mosquitoes of which our common house mosquito, *Culex pipiens*, is an example. The English sparrows, and certain other birds, in many localities are frequently infected with malaria. From them the culicine mosquitoes derive their infections, and, no doubt, every summer thousands of sporozoites of bird malaria are pumped into us by these common mosquitoes but we never become infected. There is thus no danger of becoming a martyr to science while working in the laboratory with bird malaria.

The importance of laboratory studies of bird malaria can hardly be over-emphasized. Already many important discoveries have been reported and many more should follow considering the large number of investigators now engaged in this type of work. While working with bird malaria, MacCallum discovered that the phenomenon of exflagellation that had puzzled malariologists for many years was part of the process of sexual reproduction. Ross discovered the mosquito transmission of malaria while working in India with bird malaria. More recently several important drugs that were first tested on birds infected with malaria have been found to be equally effective against human malaria. Besides these discoveries, a large mass of data of a purely scientific nature has been accumulated which may at any time be translated into economic terms for the benefit of mankind.

One would think that with hundreds of scientists studying malaria continuously year after year we would soon know all there is to be known about the subject, but, as usual with scientific work, each problem when investigated gives rise to other problems, and the completion of the work recedes always further and further into the distant future. At any rate our successors need not be afraid that there will be no problems to work on. Fortunately for us, the life history, transmission, and relations of bird malaria organisms are so nearly like those of human malaria that we can study them with every confidence that what we learn can be applied almost directly to the human type of infection.

COURSE OF INFECTION WITH BIRD MALARIA

The life cycles of the bird malaria organisms are similar to those in man. Not more than fifteen years ago all bird malaria parasites were supposed to belong to one species, but investigators have since then separated them into ten or more species. These differ from one another in their morphology and life cycles much as do the human species. No group of birds appears to be particularly susceptible, unless it is the sparrow family, since infections have been reported from about one hundred species belonging to about twenty families. For many years a strain isolated from an English sparrow in New York in August 1913 was used by most of the investigators in this country, but during the past decade a number of other species and strains have been isolated and studied from birds both in the United States and in foreign countries.

Bird malaria can be studied in the laboratory because canaries are available as experimental animals. Female canaries are not expensive; the birds are reared in captivity and hence are almost certain to be uninfected when purchased; and the malaria organisms from many wild birds will flourish in canaries. Blood is drawn up from a leg vein into a small syringe containing a little citrate solution, which prevents coagulation, and injected into the breast muscle or into a vein in the leg of a clean bird. The course of an infection with the species that has been used most extensively is usually somewhat as follows. Five days after the infective blood is injected into a bird, parasites can be found in the blood of this bird by microscopic examination. The parasites increase in number daily for the succeeding five days until they reach a peak. They then decrease in numbers about as rapidly as they increased and at the end of another period of five days not enough remain in the blood to be found by routine examination. If one examines the blood films long enough he may find a parasite, but from this time on the infection is said to be latent and this latent period may continue indefinitely. Whether or not such a bird is still infected may be determined by inoculating some of its blood into a clean bird; if this bird comes down with an infection latency still exists, but if not, the chances are that the donner had lost her infection entirely.

During the course of the infection a battle is going on between host and parasite. At first the parasite has the better of the struggle since its numbers increase steadily for about ten days. Even during this period, however, death occurs among the parasites, since statistical studies have shown that only about one-half of the young merozoites produced by each parasite at the end of a cycle of asexual reproduction succeed in penetrating and developing in fresh red blood corpuscles. The destruction of parasites becomes much greater after the peak in parasite number has been reached; and during the latent period, just enough remain alive to keep the infection from dying out. Attempts to obtain an infection in such a bird that will behave like the original infection are of no avail since the bird has acquired immunity to the organisms as a result of the first infection. This immunity seems to be associated with the activities of those scavenger cells of the body known as phagocytes which roam about seeking to devour any foreigners that may have invaded their native land. However, immunity to one species of bird malaria does not protect the bird against infection with another species, but, just as human beings may be infected with all three human malaria parasites at the same time, so canary birds may be supporting several different species with no apparent antagonism among them.

Birds exhibit symptoms during the period when the number of parasites is high; instead of hopping about they remain quietly on a perch with their feathers ruffled up. Unfortunately they can't tell us how they feel and only with the greatest difficulty can their temperature be taken, consequently we know practically nothing about the symptomatology of bird malaria. Even if we did, that knowledge would probably help us very little in our treatment of human malaria, since we know all that is necessary regarding human symptoms. However, many other phases of the disease in birds may throw light on human malaria.

PROBLEMS IN BIRD MALARIA

For many years one of the unsolved problems in malaria was that of the relations of the parasites to the red blood corpuscles. Certain investigators claimed that the parasites cling to the outside of the corpuscle, and that, during their growth, migrate from one corpuscle to another. Thus a single parasite might destroy several corpuscles. Eventually a method of settling this problem was devised. The blood of parasitized birds was embedded in paraffin and cut into very thin sections. Some of the red blood corpuscles were sectioned in such a way that the parasites could be seen inside of the corpuscle surrounded by the outside wall. In this way it was proved that the parasite is "inside looking out" instead of "outside looking in." There is of course, still the possibility that some parasites cling to the outside of the corpuscles, but the probability that this occurs seems rather remote.

Another problem of considerable interest was concerned with the activities of the parasites during the period of latency. Where are they located and how do they reproduce at this time? Long and tedious studies of bird malaria seem to have solved this problem also. Birds were killed at various intervals during the latent period and thin sections of tissue scrutinized with care. Occasionally a parasite was found, and when all the evidence was collected, the conclusion was reached that the parasites are largely hidden in the spleen and bone marrow and that they maintain themselves there by asexual reproduction, which does not differ from reproduction during the acute stages of the infection except that most of the offspring are destroyed. Students of free-living protozoa find this situation quite remarkable. For example, a strain of the common ciliate paramecium can be maintained apparently indefinitely by asexual reproduction if the medium in which it lives is changed frequently, but will die out in an environment that is kept constant. One investigator actually carried on a culture of paramecia in this way for twenty-five years during which about 15,300 asexual generations were noted, but no sexual reproduction (conjugation) occurred. This has been considered rather an extraordinary feat. The paramecia, however, had to be watched very carefully to prevent conjugation, and furthermore were known to pass through a sort of internal reorganization (endomixis) at intervals,—a process that appears to restore the vitality of the strain.

On the other hand, malaria organisms reproduce only by the

asexual method while in the vertebrate host; they live at a constant temperature and in a constant medium, the blood; and they do not undergo any internal reorganization, so far as is known, as paramecia do. When transferred from one bird to another by blood inoculation, the intermediate host (mosquito) in which the sexual stages in the life history occur is entirely eliminated. Therefore, the strain of bird malaria which was isolated from an English sparrow in 1924 and has been passed from canary to canary by blood inoculation ever since then, and which, furthermore, reproduces asexually every twenty-four hours, must have survived without the intervention of sexual reproduction for thirteen years or over 4,000 generations. No evidence of loss of vitality has been noted in this strain, and it is safe to conclude that bird malaria organisms can exist indefinitely in a constant medium without sexual reproduction or internal reorganization and without any loss in vitality.

PERIODICITY IN MALARIA

Periodicity in the appearance of symptoms is one of the characteristics that distinguishes malaria fever from certain other types of fever. Soon after the true organism of malaria was discovered a definite relation was observed between the symptoms and the course of the infection. The asexual cycle was found to be definite in length; most of the parasites were in the same stage of development at the same time, consequently a synchronous breaking down of the red blood corpuscles occurred when a new generation of merozoites escaped into the blood; and the host experienced chills and fever at definite periods corresponding to the simultaneous destruction of the red blood corpuscles. We now use the term periodicity to indicate the length of successive asexual cycles. For example, in human malaria the tertian parasites pass through an asexual cycle in 48 hours, the quartan parasites in 72 hours and the aestivo-autumnal parasites in 24 to 48 hours. Periodicity in the tertian and quartan infections is quite constant and synchronicity is high, that is, most of the parasites reproduce at the same time. Periodicity and synchronicity are less well-marked in infections with the aestivo-autumnal parasite.

Similar results have been obtained as regards periodicity and synchronicity from studies on bird malaria. The ease with which birds can be subjected to modifications, however, render them more satisfactory than man for experimental purposes. Establishing the character of the periodicity is relatively simple; determining the factors involved is difficult. That the environment of the host has a profound effect on the parasites is indicated by recently reported experiments. For example, the species of bird malaria parasite (*Plasmodium cathemerium*) that has been employed most extensively for experimental studies, undergoes a daily asexual cycle and the new generation breaks out of the red blood corpuscles during the evening hours (5 to 9 P.M.). If, however, parasitized canaries are kept in the dark during the day and in a lighted room during the night, the time of sporulation changes within three days to the corresponding morning hours (5 to 9 A.M.). If clean birds are then inoculated with both kinds of parasites, that is, with some that sporulate in the evening and others that sporulate in the morning, and are kept under normal conditions, those that sporulate in the evening continue to do so, and those that had been sporulating in the morning change back to the evening hours. When infected birds are exposed to the light for fourteen hours and then to darkness for fourteen hours for a few days the asexual cycle is increased from 24 to 28 hours.

These experiments suggest that light regulates periodicity in bird malaria, but further studies indicate that light may not be of primary importance; neither is fatigue, which was one of the factors studied; but some influence related to the ingestion of food now seems to control this phenomenon. We have already discussed the conspicuous effects of changes in diet on intestinal protozoa and it seems rather strange to find that food may also exert an influence on blood-inhabiting protozoa. But digested food is taken up by the blood stream and might easily affect protozoa living in that habitat.

EFFECTS OF SUGAR ON BIRD MALARIA

Several years ago we attempted to modify bird malaria infections by changing the sugar content of the blood. The clues that led to our experiments came from several sources. Bass, in 1912, found that the addition of sugar to his culture medium was necessary before he was able to cultivate malaria parasites successfully outside of the body. In 1913, Bass and Johns

FIG. 114. Who can blame malarial parasites for demanding the American standard of living—well sweetened.

reported the cultivation of the organisms of human malaria in the blood of a diabetic without the addition of sugar. Certain drugs, such as epinephrin, that bring about an increase in the number of parasites in the blood, and are known as provocatives, are also effective in increasing the sugar content of the blood. These facts suggested that the blood stream can be improved as a culture medium for the growth and multiplication of malarial parasites by the addition of sugar. According to this theory, raising the sugar content of the blood would result in an increase in the number of parasites. *Vice versa,* the blood stream should be rendered less favorable than normal as a habitat for malarial organisms by lowering the sugar content. Glucose was fed to the birds in order to raise the sugar content and insulin was used in order to lower it.

The details of our experiments need not be described here. In general the results indicate that an increase in sugar content renders the blood more favorable for the malaria parasites; the number of parasites becomes greater and remains so for a period longer than normal, relapses may occur earlier in the course of the infection and more frequently than normal; and a larger proportion of the birds die as a result of the infection. When treated with insulin the parasite curve is lower than normal throughout the course of the infection. So far as we know, no attempt has been made to modify refractory foreign hosts in this way so that they may become more susceptible, but it seems to us probable that positive results would follow experiments of this nature. For example, chickens which can not be infected with the parasites of sparrows might become less resistant if they were fed on glucose.

EFFECTS OF TEMPERATURE AND HUMIDITY ON BIRD MALARIA

Another method of modifying the host that has been used in the study of malaria is to change the atmospheric humidity and temperature. Malariologists have found that an increase in these factors favors the breeding of mosquitoes and the development of sporozoites within the mosquito, and hence is responsible, at least in part, for epidemics of malaria. We attempted to determine whether such changes modify canaries so as to affect the development of the asexual and sexual stages in the blood. Birds were inoculated with infected blood and maintained under normal conditions in a room where the temperature was about 70°F and the relative humidity about 40 per cent. Other birds were kept in a special room at a temperature of about 100°F and a relative humidity of about 90 per cent. The course of the infections in these birds was carefully followed. The data obtained show that the average number of sexual and asexual parasites was higher in the experimental birds than in the controls and that both types of parasites were present in the blood in larger numbers for a longer period. Thus the higher temperature and humidity appeared to modify the host so that the parasites were able to set up a more severe infection extending over a longer period of time.

PROBLEMS FOR THE FUTURE

As noted above, birds are being more widely used for malaria studies than ever before. Some of the problems in bird malaria that seem to us most important at this time are of immediate practical value and others of purely scientific interest, but of such a nature as to suggest practical applications. One of the outstanding problems is that of securing a drug that will actually destroy all of the parasites in the body and not leave a few alive to keep up the infection with. Another desideratum that is greatly needed is a method of determining whether or not parasites are present when there are too few to be found in the blood by microscopic examination. In birds we can determine this point by inoculating a clean bird with the suspect's blood, but must wait five days or longer for a result. Probably some serological technique will eventually be devised of use both in bird and in human malaria.

The battle between parasite and host involves several operations about which much more needs to be learned. For example, we know that the malaria organisms destroy red blood corpuscles but don't know whether they also elaborate toxic substances that are harmful to the host, and if so, what these toxic substances are. Why one species or strain of malaria parasites is more virulent than others is likewise an unsolved problem. Furthermore, we need to know more about how the host destroys the parasite. Evidently the phagocytic cells play a prominent rôle in this slaughter, but other types of resistance of a more subtle nature may be destructive to the enemy forces. Very little is known about the acquisition of immunity to malaria as a result of infection extending over a long period, but such an immunity seems to develop in natives living in malarious regions. The mechanism of this immunity requires further investigation.

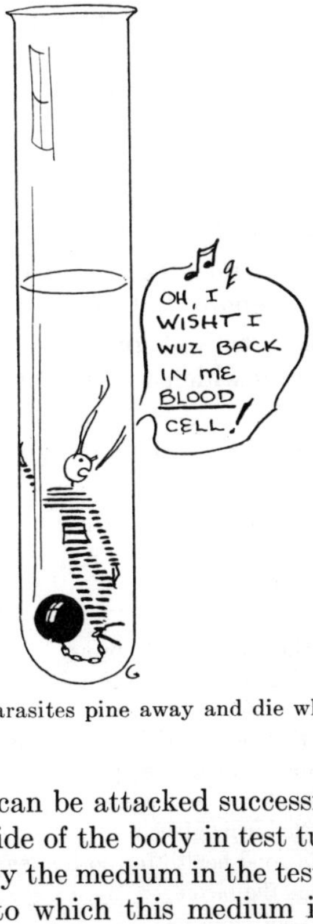

FIG. 115. Malarial parasites pine away and die when imprisoned in a test tube.

Many problems can be attacked successfully when protozoa are cultivated outside of the body in test tubes. The environment represented by the medium in the test tube and the light, temperature, etc. to which this medium is subjected, can be modified in various ways and the effects on the organisms observed. Most of the human protozoa have been cultivated in artificial media; these include malaria parasites. However, the methods of cultivating malaria parasites thus far devised have not led to any results of value. Improvements in these methods are, therefore, very much to be desired. Since human malaria parasites were first cultivated in vitro (Bass, 1911), no one has succeeded in maintaining cultures in a viable condition for more than a few days. We hope for better luck with the organisms of bird malaria if and when some investigator seriously undertakes this problem.

One disadvantage we must contend with when we use canary birds as winged guinea-pigs is the small amount of blood contained in them which renders serological work difficult. Recently a species of malaria that will live in domestic fowls has been obtained in the Far East and imported into France. We would like very much to work with this species, but, due to government regulations, are prevented from importing the organisms into the United States. However, enough problems still remain that can be studied in canaries to keep us busy for many years to come, and so we are now living in that enviable state in which we are never at a loss to know what to do next.

Gilbert and Sullivan wrote their delightful operettas before the malaria organisms of either man or bird were known, otherwise they might have made some of their songs instructive as well as entertaining. The following is an example of what might have been.

The birdies that sing in the spring. Tra la!
Have something to do with the case.
We prickle them under the wings. Tra la!
And pepper their blood with such things. Tra la!
That fever and chills thrive apace.
Oh! aren't we all red in the face.
And then little birdie gives us a hot clue
That leads to good health for both me and for you.
Tra la! Sings the birdie;
Goodbye to your chills;
The ague has left you at last.
Tra la! Sings the birdie;
Goodbye to dull care;
Quinine and fever are past.

Chapter X

CONQUISTADORES

IF THE previous chapters in this book have accomplished their purpose you should by this time possess a somewhat definite idea of the host-parasite relations that exist between man and some of his protozoan parasites. But, like the small boy who eats his cake first leaving the frosting until the last, we have reserved the most important of all protozoan diseases for the concluding chapter. Everyone recognizes the personal and public-health significance of such diseases as smallpox, diptheria, cancer and tuberculosis, but in the north temperate zone one does not acquire a real appreciation of the devastating force of malaria. Not until we plot on a map the distribution of this disease, consider the number of human beings who are attacked by it year after year and contemplate its economic seriousness do we obtain an adequate idea of malaria as a world problem.

SERIOUSNESS OF MALARIA

In certain regions covering thousands of square miles in the warmer parts of the world everyone is infected with malaria from the time he is a few days old until he dies. When a child is to be born, malaria mosquitoes are already present before the doctor arrives, avidly awaiting an opportunity to greet the newcomer with a dose of malaria germs. If the baby survives and grows up, he never knows for a single day what it is like to be well. If he does not die of malaria when he is a child, or of malaria in combination with some other disease, he acquires a certain amount of immunity and can carry on the activities of life more or less inefficiently. But he may at any time suffer an acute attack which transforms him from a working man into a helpless invalid. Such people may well be pessimistic; for example, one old man, who was asked when he had had his last attack of malaria, answered, "Not yet" (Hackett).

The suffering and loss of life due annually to malaria is greater than that of any war that was ever fought. It is even greater than the mortality due to automobile accidents. Economically malaria is equally serious since it alone prevents the colonization of vast areas of the most fertile land on the earth's surface. How remarkable that this should be true, since we

FIG. 116. In some regions mosquitoes squirt the babies full of malaria germs as soon as they are born.

know the organisms responsible for the disease and can easily recognize them; have a detailed knowledge of their method of transmission and have devised effective methods of control.

PARASITES OF HUMAN MALARIA

Three species of the genus *Plasmodium* are the parasites of human malaria. These are *P. vivax*, the organism of benign tertian malaria; *P. malariae*, the organism of quartan malaria, and *P. falciparum*, the organism of estivo-autumnal or malignant tertian malaria. Malarial fevers are characterized by periodicity of the symptoms, by the deposition of pigment in the tissues of the body, by anemia, and by enlargement of the spleen. Various synonyms of the term malaria exist in English and other languages, such as ague, marsh fever, and intermittent fever in English, paludisme and fievre palustre in French, paludismo in Italian, and Wechselfieber in German.

Benign tertian malaria occurs throughout most of the area shown on the malaria map, but is more prevalent in the cooler portions of its range. Patients suffering from this type of malaria experience chills and fever every forty-eight hours, that is, on every third day, hence the name tertian. The attacks of this species of parasite are not as severe as those of *Plasmodium falciparum,* hence the disease they provoke is called benign tertian in contrast to that of the latter which is known as malignant tertian malaria. This type of malaria is more frequent in the hotter portions of the world. It likewise is tertian in its symptoms. Quartan malaria, which is characterized by chills and fever every fourth day, is not very common anywhere and is ordinarily not considered very important when compared with the other two types.

HISTORY OF MALARIA

Probably no other disease has ever attracted as much attention or stimulated the formulation of so many exploded hypotheses. Again and again conclusions were reached by competent scientists regarding the cause of malaria and its method of transmission, only to be disproved. Today it seems almost incredible that these problems were not solved before 1880, when the true protozoan parasite was discovered, and before 1898, when the mosquito transmission of the parasite was fully established.

Torti, in 1753, seems to have been the first to use the term malaria,—a name consisting of the two Italian words *mal* and *aria,* meaning bad air. This term was quite appropriate at that time since the disease was thought then and even much later to be due to some evil characteristic of the air. Earlier physicians confused malaria fevers more or less with other fevers. However, the disease was well known to the Greeks at least 2500 years ago and quotidian, tertian, and quartan fevers recognized. Hippocrates mentions one of its chief symptoms, the enlargement of the spleen, and noticed the reduction in the size of this organ when the patient recovered.

Many curious ideas were expressed to account for the periodicity of the chills and fever. Darwin suggested the periodical recurrence of sleeping and waking; Werlhof, the periodical movement of the earth; Mead and others the influence of the moon; and Bailly, the changes in the position of the human body, from the upright by day to the recumbent by night and *vice versa.* The disease was by Oldham considered to be due to a sudden chill such as might result from the dampness prevalent in marshy regions. Munro started a polemic with Oldham, claiming priority for this discovery.

Many writers attributed malaria to gases emanating from marshes, such as sulphuretted hydrogen. Others believed that organic matter was involved. Bence-Jones states that "Ague ferment is probably a highly complex nitrogenous substance, capable of being dried and carried by the wind far from the place where it was produced. It enters by the mouth with the dust, and, like animal or vegetable alkaloids, it passes from the blood into every texture of the body, and acts on each much or little, according to its chemical properties." This author's reference to the carriage of the ferment by the wind was no doubt an attempt to explain the rôle of air currents in the transmission of malaria. We now know that the virus may be carried by the wind but only when it is inside of an infected mosquito.

Wind transmission, however, was never as popular as water transmission, and even today there are thousands of people who believe that malaria is due to bad drinking water.

That malaria originated in the soil was another favorite theory. For example, Sternberg in 1884, which was 4 years after the discovery of the real malaria organism, writes as follows: "The cause of the periodic fevers—malaria—is of telluric origin. The evidence in favor of this assertion would fill volumes, and is beyond question. Those who go to sea in ships escape these fevers, in the latitudes where they prevail, until they approach the land. In tropical regions those who remain on board a vessel anchored at some distance from the shore remain in good health, while the boat-parties, sent ashore for provisions, etc., fall sick with the fevers of the country. We have African fevers, Panama fever, Bengal fever, etc., but who ever heard of Indian Ocean fevers, or Mediterranean fevers or Gulf fevers? Any explanation, then, which attributes these fevers to general causes, the action of which is independent of locality, must be rejected."

"Malaria, although indisputably of telluric origin, is not given off from every kind of earth. Those who mould clay into bricks or earthen-ware, and those who excavate banks of pure sand or gravel, are not subject to attacks of malarial fevers by reason of their occupation. On the other hand, agricultural laborers who are exposed to the emanations from a soil rich in organic material are especially liable to be attacked by these fevers.

The evidence connecting the production of malaria with the presence of organic matter in the soil is overwhelming and conclusive. The general rule, however, holds good that soils rich in organic matter are most prolific of malaria. Such soils are found in low, marshy places, in the deltas of great rivers, in the broad alluvial plains bordering great rivers, and in the valleys of smaller streams."

DISCOVERY OF THE MALARIA ORGANISM

A number of investigators discovered living organisms which they claimed to be the agents of malaria before the true parasite was revealed by Laveran. Thus Salisbury in 1866 concluded that a unicellular alga, a microscopic plant, was the malarial germ. Another species of alga that was abundant in the Pontine Marshes near Rome was implicated by Bolestra in 1880. Eklund of Sweden in 1878 not only claimed that the spores of a fungus were responsible for malaria but found these spores in the decaying vegetation in marshes and in the blood of patients. The organism, however, that was believed by most people at that time to be the malaria germ was a bacterium called by Klebs (1879) *Bacillus malariae.* Klebs' work was confirmed by various investigators and so firmly entrenched was the belief in this organism and so frequently had the claims of scientists been disproved that when Laveran in 1880 described the true parasite very few paid any attention to the discovery. We thus see that scientists from many nations

FIG. 117. Many curious ideas were expressed regarding the cause of malaria before the malarial protozoon and mosquito vector were discovered.

took a shot at this perplexing problem. And all the time mosquitoes full of the real malaria organisms were sitting just around the corner enjoying the vociferous arguments of these solemn and deluded scientists.

Every great discovery is usually preceded by a series of minor observations that provide suggestions. Thus Meckel in 1847 noted pigment granules in the body of a woman who had died of malaria, and Virchow in 1848 actually observed pigmented bodies in the blood, which later were recognized as diagnostic of malaria. When Laveran began the studies that led to his great discovery he decided to investigate the nature and origin of the pigmented elements in the blood. These he reported in 1880 to be the parasites of malaria, which at first he thought were algae belonging to the genus *Oscillaria*, but later decided were protozoa.

Laveran, in the introduction to his treatise on malaria fevers, published in 1884, relates the story of his discovery as follows:

"This work is the result of five years of research pursued in Algeria without relaxation. In writing it, the principal aim has been to make known the parasites which I have had the honor of discovering in the blood of patients afflicted with malaria fever.

On my arrival at Bone in the month of September, 1878, I did not nourish the hope of discovering the cause of malaria and my first efforts were not made in this direction. I occupied myself entirely at first with the clinical side and pathological anatomy of malaria fevers, striving to free myself from all preconceived ideas, having made a clean sweep of the classical theories." This attitude was probably largely responsible for Laveran's discovery, since as Darwin has pointed out, "The truth will not penetrate a preoccupied mind." Laveran continues as follows:

"The analysis of anatomical lesions observed on subjects dead of pernicious malaria made me realize that the characteristic lesion, the only constant one in malaria, consisted in the presence of pigmented elements in the blood. These pigmented elements were already known, but their nature and their origin were still obscure. It was in studying the mode of formation of these pigmented elements in the blood that I was brought to recognize their parasitic nature.

On the 6th of November 1880, I examined the blood of a patient who was being treated for intermittent fever at the Military Hospital at Constantine. Here I observed for the first time the existence of mobile filaments which adhered to the pigmented bodies, and the living nature of which was undoubted. I had at this moment, even, the intuition that I

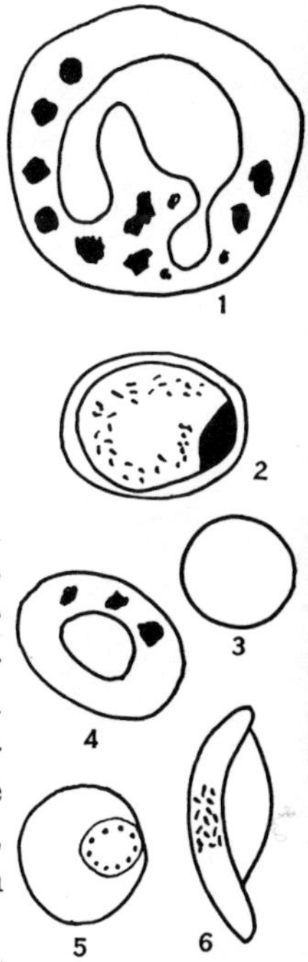

1. Phagocyte containing malarial pigment.
2. Femal sexual cell of quartan malaria with many pigment granules.
3. Normal red blood corpuscle.
4. Phagocyte containing pigment. 5. Red blood corpuscle containing a parasite.
6. Crescent-shaped sexual cell. (Figures 3, 4, 5 and 6 were copied from Laveran's drawings of 1884.)

was in the presence of the true microbes of malaria and all the facts that I have observed since then have only confirmed this first impression." The "mobile filaments" that Laveran saw were, of course, microgametes still attached to a microgametocyte.

About ten years were required to confirm Laveran's discovery, to learn to distinguish one species of malaria parasite from another and to work out the life-cycles of the parasites in man. In the meantime this discovery stimulated intensely attempts to determine the method of transmission.

METHOD OF TRANSMISSION OF MALARIA

Many previous researches paved the way for the discovery of the mosquito method of transmission. That parasites may normally live a double life during which they may occupy two hosts of different species was discovered by Abildgaard as long ago as 1790. This investigator found that a certain type of tapeworm (*Bothriocephalus*) spends part of its life in fish and the other part in water-fowl that feed on these fish. This was followed some years later by similar revelations in the life cycles of other parasitic worms. Then Leuckart and Melnikoff discovered a case in which an insect, the dog louse, was found to be the intermediate host of the dog tapeworm, *Dipylidium caninum*.

The parasitism of insects by protozoa was also known long before the mosquito-transmission of malaria was discovered. For example, Dufour in 1828 found many different species of gregarines in insects and Leidy in 1849 recorded ciliates from the cockroach and in 1879 described parasitic amebas from the same host. Of particular importance was the announcement by Smith and Kilborne in 1893 that the protozoan parasite of Texas fever is transmitted from diseased cattle to healthy cattle by a tick.

Numerous hypotheses regarding the method of transmission of the malaria parasite were made involving the mosquito before Ross proved in 1898 that this insect is the intermediate host. Of these hypotheses Ross writes: "It is easy to sit at home and weave many hypotheses; but attempts to verify them generally demand endless labour, expense, self-sacrifice or even danger, and often fail, or, even if successful, win little reward. There were many before Columbus who imagined America; but between the dream and the reality an ocean had to be traversed."

Ross began his mosquito studies in May 1895 when he fed *Culex* and *Stegomyia* mosquitoes, bred from larvae, on infected human beings. He noted exflagellation in the stomachs of the

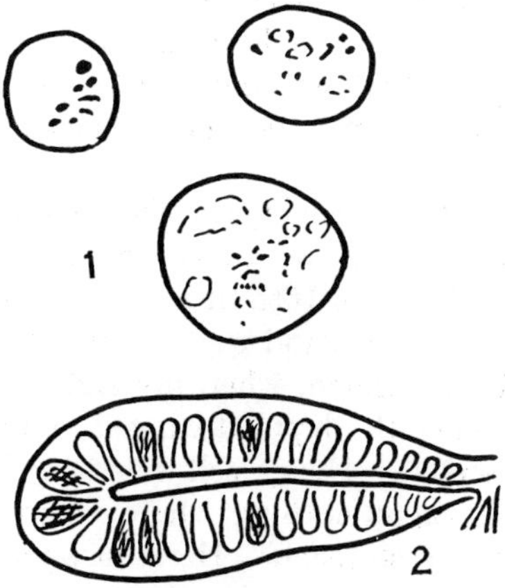

FIG. 118. Copies of drawings made by Sir Ronald Ross. 1. Three oöcysts from the stomach wall of anopheles mosquitoes (Aug. 20, 1897). 2. Part of a salivary gland containing sporozoites (July 18, 1898).

insects but could not follow them further. In 1896 he fed species of these two genera of mosquitoes on an infected person and then on a healthy person, of course, without results, since he had selected the wrong species of mosquito. Ross realized that perhaps he was not working with the right species, so in August 1897 he fed dapple-winged (anopheline) mosquitoes on an infected person and afterward found oöcysts in two of them. He was unable to continue experiments on human malaria because he was ordered by the British government to a non-malarious region, so he undertook a study of bird malaria.

In March 1898 he fed *Culex fatigans* on birds containing malaria parasites and at once found oöcysts. Later (July 4, 1898) he located sporozoites in the salivary glands of a mosquito and in July and August infected 22 of 28 healthy sparrows by means of the bites of infected mosquitoes (*Culex fatigans*). Only female mosquitoes bite animals hence only females can transmit the disease. Thus was established for all time, even without the aid of Kipling, the well-known fact that the female of the species is more deadly than the male. The male may have more whiskers, represented by the hairs on his antennae, but the female is a ghoul with thin wings and streaming legs, sounding her high-pitched, hateful bugle in the air or vibrating with obscene ecstasy as she sucks our life blood like a midnight vampire.

Ross described his discovery in the following words: "On August 20, 1897, I had two remaining insects, both living. Both had been fed on the 16th instant. I had much work to do with other mosquitoes, and was not able to attend to these until later in the afternoon, when my sight had become very fatigued. The seventh dappled-winged mosquito was then successfully dissected. Every cell was searched, and to my intense disappointment nothing whatever was found, until I came to the insect's stomach. Here, however, just as I was about to abandon the examination, I saw a very delicate circular cell, apparently lying amongst the ordinary cells of the organ, and scarcely distinguishable from them. Almost instinctively I felt that here was something new. On looking further, another and another similar object presented itself. I now focussed the lens carefully on one of these, and found that it contained a few minute granules of some black substance, exactly like the pigment of the parasite of malaria. I counted altogether twelve of these cells in the insect, but was so tired with work, and had been so often disappointed before, that I did not at the moment recognize the value of the observation. After mounting the preparation I went home and slept for nearly an hour. On waking, my first thought was that the problem was solved, and so it was."

Then Ross proceeds to add: "The mind long engaged with a single problem often acquires a kind of prophetic insight, ap-

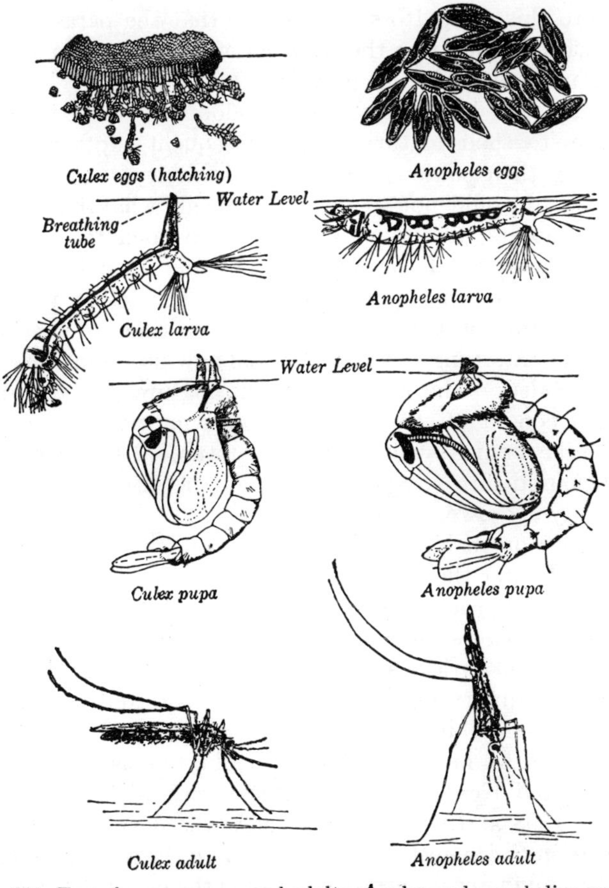

FIG. 119. Eggs, larvae, pupae, and adults of culex and anopheline mosquitoes, illustrating differences in structure and habits.

parently stronger than reason, which tells the truth, though the actual arguments may look feeble enough when put upon paper. Such an insight is mainly based, I suppose, on a concentration of small probabilities, each of which may have little weight of itself; but in this case, at all events, the insight was there, and spoke the truth."

Up to this point Ross had proved that the parasite of malaria when sucked into the stomach of mosquitoes in a meal of blood underwent further development.

"I had traced," he says, "the development of the pigmented cells up to their maturity and subsequent rupture and discharge of their contents into the body cavity of the grey mosquitoes. I could not see at the moment what happened to these contents, yet upon this point depended the vastly important question of the route of infection in malaria. . . ."

"Hitherto my mosquitoes had been dissected in water or a weak solution of salt, and I had had no time for methodical staining. A strong salt solution was now used, and the secret was revealed. The contents of the mature pigmented cells did not consist of clear fluid, but of a multitude of delicate, threadlike bodies, which, on the rupture of the parent cell, were poured into the body cavity of the insect. They were evidently spores."

Further on in his narrative he shows how the spores here described enter the salivary gland of the mosquito.

"The exact route of infection of this great disease, which annually slays its millions of human beings, and keeps whole continents in darkness, was revealed. These minute spores enter the salivary gland of the mosquitoes, and pass with its poisonous saliva directly into the blood of man. Never in our dreams had we imagined so wonderful a tale as this."

Koch and Kessel confirmed Ross' work with birds in September 1898 and were able to follow the development of the parasite in *Culex nemorosus.* They were the first to determine how the oökinetes penetrate the wall of the insect's stomach.

After Ross had proved the transmission of bird malaria by mosquitoes certain Italian investigators, especially Grassi, were able to work out the transmission of human malaria. In November 1898 Bignami infected a man in Rome with mosquitoes, and a few weeks later Bignami, Bastianelli and

Grassi found the oökinetes of malignant tertian malaria in *Anopheles maculipennis* and also proved that benign tertian malaria is carried by anophelines.

The final proof of the mosquito theory was obtained with the aid of Dr. P. T. Manson of London. Ten anopheline mosquitoes were allowed to infect themselves by biting a patient in Italy who had benign tertian parasites in his blood. These

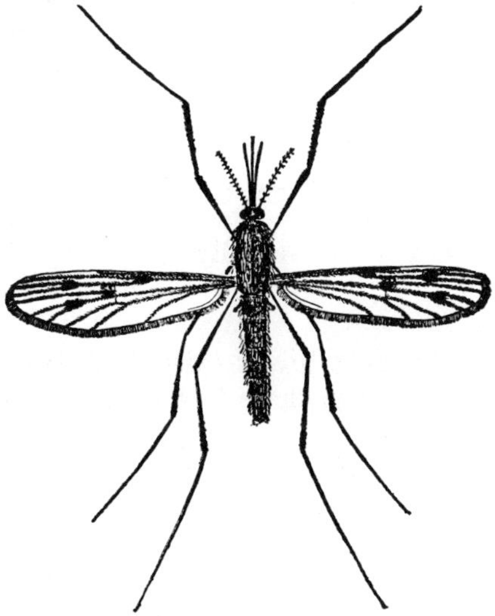

FIG. 120. A malaria-carrying mosquito, *Anopheles quadrimaculatus*

mosquitoes were then sent to London and Dr. Manson permitted himself to be bitten by them between August 29 and September 4, 1900. After feeling perfectly well until September 13, a severe headache, sensations of chilliness, lassitude, pains in the back and bones, and a temperature of 101.4°F. developed. These premonitory symptoms were followed on succeeding days by chills and fever. Parasites were found in his blood on September 16 and became numerous on September 17. Quinine was then administered and the course of the disease arrested.

MALARIA AMONG THE ANCIENT GREEKS AND ROMANS

Although several books have been written on the subject, no one really knows how great an influence malaria had on ancient Greek and Roman civilizations. It seems hardly probable that Greece could have been malarious during the height of her prosperity considering the effects of this disease on vigor, both physical and mental. According to one suggestion, malaria was introduced into Greece during the fifth century B.C. by soldiers returning from foreign wars, or by slaves. The disease would have been severe in such a non-immune population and would soon have undermined any civilization no matter how advanced, if the cause and methods of control were not known. The Greeks celebrated their foreign conquests with great pomp and circumstance, little realizing that they were actually enslaving themselves to malaria parasites that were destined eventually to destroy them.

The Greeks knew that fevers existed of the malaria type, and were also aware that these fevers were prevalent in the neighborhood of marshes. Thus Empedocles "delivered Selinus (in Sicily) from a plague by draining its marshes," and Hippocrates notes that the inhabitants of low, meadowy and hot areas were below normal in health and intellect. Plutarch in describing the death of Pericles mentions symptoms very suggestive of malaria, and Pausanias tells us that the deterioration of the Greeks in the third century B.C. was in part due to diseases, which no doubt included malaria. When a tug of war is properly conducted, one side or the other always wins; in this case the Greeks were easily defeated by the malaria mosquito.

Similar arguments have been assembled in favor of the idea that the decline and fall of the Roman Empire was also largely due to malaria. Many ancient Roman cities which are now highly malarious were probably built at a time when the surrounding regions were free from malaria. The later introduction of the disease would account for their abandonment.

MALARIA IN THE NEW WORLD

Whether malaria existed in the New World in pre-Columbian times or was introduced by the Spanish Conquistadores is uncertain, but we know that the first city to be established, namely Isabella in Hispaniola (now Haiti), which was founded by Columbus in 1494, was abandoned because of epidemic diseases. Later, the members of the expedition led by Balboa across the Isthmus of Panama suffered severely from fevers, as did also the Spaniards who attempted to colonize Darien. Balboa got a thrill when he discovered the Pacific,

FIG. 121. The tug of war between the Greeks and Malaria mosquitoes was easily won by the mosquitoes.

but the mosquitoes must have been even more thrilled when they discovered Balboa.

In Peru, the medicine men of the ancient Incas appeared to be familiar with the malarial type of fever and had discovered the value of cinchona bark in its treatment. This discovery was one of the greatest in the history of medicine. The Countess of Chinchon, who lived in Lima, Peru, was cured of an attack of benign tertian malaria by means of cinchona bark in 1638. On her return to Spain in 1640 she took with her a supply of this bark thus introducing to the rest of the world the raw material from which quinine was later manufactured.

Further north the English were having trouble on account of diseases in establishing colonies. For example, at Jamestown in 1607, fifty of one-hundred and five colonists died so rapidly between May and September that the "living were scarce able to bury the dead." We are not certain that these deaths were due to malaria, but Jamestown was ideally situ-

FIG. 122. While Balboa was discovering the Pacific Ocean, mosquitoes were discovering Balboa.

ated for the propagation of this disease, and was later deserted because of its unhealthy location. Even in New England the early settlers were beset with chills and fever, and epidemics of malaria are on record in Massachusetts in 1647, 1650 and 1668. Malaria was likewise one of the most difficult obstacles met by those hardy pioneers who migrated westward during the early years of the republic and settled in the Mississippi Valley. In the northern part of this region, the disease gradually disappeared as the land was brought under cultivation, since "malaria flees before the plough"; but in the southern part of the valley malaria is still prevalent.

That malaria is still one of the most important of all diseases in certain parts of the United States is indicated by the fact that in 1934 approximately 3,900 deaths due to this disease occurred in thirteen of our southern states, whereas during the same period there were only 2,200 deaths from typhoid and 2,000 from diphtheria in the same area. Certain other countries however, are much worse off than we are, and the estimates of death and destruction almost reach the realms of higher mathematics.

MALARIA IN INDIA

Malaria may be responsible for death either *per se* or indirectly by weakening the individual so that he fails to combat successfully some other disease. It is estimated that malaria is thus accountable for about 2,000,000 deaths each year in India alone. This however, is only a small proportion of the number who annually suffer from malaria in that much-afflicted country,—a number estimated at 100,000,000. Those who have seen people suffer year after year from malaria can understand fully what this means in terms of human misery.

MALARIA CONTROL IN PANAMA

Perhaps the best-known campaign that has ever been waged against any disease was that of General Gorgas for the eradication of malaria and yellow fever in the Panama Canal Zone. This triumph in the field of preventive medicine came soon after the mosquito-transmission of malaria had been discovered by Ross and Grassi. The medical officers of the U. S. Army had already freed Havana, Cuba, of these two diseases and had effectively demonstrated how malaria could be controlled under urban conditions, at the time work was begun in Panama in 1904. Here a very different situation was encountered. A strip of land forty-two miles long, through which the canal was to be dug, had to be treated so that workmen living under more or less primitive, rural conditions would be protected from infection. The method of attack was to divide the area into eighteen districts, each in charge of an inspector. Each inspector was allowed a force of about fifty men, and seven methods of control were inaugurated.

Drainage designed to deprive the malaria mosquitoes of breeding places was considered to be the most important line of attack. The methods used were such as any intelligent farmer in the United States could carry on successfully; the only special knowledge necessary was that of the breeding habits of the mosquitoes living in the area. Next in importance was the cutting down of brush and tall grass around villages, camps and isolated houses so as to eliminate the shel-

FIG. 123. The United States Army quickly drove the malaria mosquitoes beyond the limits of the Panama Canal Zone.

tering places of adult mosquitoes. The larvae of the mosquitoes were attacked directly by spreading oil on the surface of bodies of water that could not be drained easily, and by mixing a poisonous larvicide in the water. About 4,500 barrels of oil and 1,900 barrels of larvicide were used per year.

Quinine was available at all times for use as a prophylactic at the twenty-one dispensaries located at various points along the canal. No one was forced to use the drug but about one-half of the employees took a dose every day. About 3,200 pounds of quinine were dispensed per year. All Government buildings were screened so as to prevent mosquitoes from infecting those within who might be susceptible or from becoming infected as a result of ingesting blood from those suffering from malaria. Wherever mosquitoes were not excluded by screening, members of the sanitary force were detailed to search out and kill them. Those that had fed during the night and were resting after their meal in a dark corner or in some other hiding place were also destroyed so far as possible. This proved to be a very efficacious method of destroying infective mosquitoes.

These control methods resulted in a rapid decrease in morbidity and mortality from diseases transmitted by mosquitoes. Yellow fever was eradicated from the Canal Zone by the year 1906. Malaria still exists there, but the inhabitants are extraordinarily mosquito-conscious and usually call a policeman when they find one. The death rate from malaria in the total population of the Canal Zone was reduced by General Gorgas from 16.21 per thousand in July, 1906 to 2.58 per thousand in December, 1909, and among the Canal employees, from 11.59 to 1.23 per thousand during the same period. Rates of admission to hospitals due to malaria are even more impressive; among employees the rate fell from 1263 per thousand in July, 1906 to 191 per thousand in December, 1909. Besides this, the Canal was dug.

MALARIA CONTROL IN THE PONTINE MARSHES

Many attempts have been made to reclaim the Pontine Marshes, or Agro Pontino, which is a large plain in Central Italy extending for a length of about thirty miles from Rome to Terracina, and ranging from five to eighteen miles in breadth. The Via Appia forms a sort of axis through this area. At one time the entire western coastal region of Italy was thickly populated. Then the Goths, and afterward the Saracens, ravaged the country and what was left was destroyed by the armies of Pisa and Siena. These armies left behind them a rear guard of malaria to "mop up" any human beings who might have escaped their military activities. Thousands of acres of productive land were thereby made uninhabitable for mankind and only water-buffaloes, horses, sheep, cattle, and mosquitoes could exist there.

Goethe crossed the Pontine Marshes in 1787, which no doubt accounts for the words of Faust, who says, "A marsh spreads at the mountain's foot, and poisons all that has already been won. To drain that pestilential pen would be the last and greatest victory. Wide lands I would open up for millions upon millions of men, lands where they could dwell . . . amid green fruitful fields . . . Would I might see that teeming host, and stand amid a free people in a free land." Many attempts were made to win this victory, by Pope Leo X at the beginning of the sixteenth century, by Sixtus V at the end of the same century, by Pius VI in 1777, by Pius VII in 1817 and by many others, but every attack ended in miserable failure. Up to 1930 "not a single permanent habitation existed on the Pontine Marshes," "and every year between June and September the mud and wattle huts of the swineherds and charcoal-burners remained deserted as the migratory occupants abandoned the fever-stricken area and returned to their mountain villages."

The latest assault appears to be bringing this longest war in history to a glorious victory for the malariologists. The five fundamental lines of attack recognized as the most effective at present have been pushed with vigor and all at the same time. (1) Infected human beings have been treated so as to reduce to a minimum the number of gametocytes in their blood that might be infective to mosquitoes. (2) Dwellings have been made mosquito-proof, so that susceptible mosquitoes cannot enter and become infected by biting human beings who carry gametocytes in their blood. (3) Mosquito larvae have been destroyed in their breeding places by spreading oil upon the waters and by dusting Paris green upon the surface. (4) Lagoons, canals and ditches have been treated so as to prevent the breeding of mosquitoes, and other breeding places have been eliminated permanently. (5) Natural methods of control, such as the eradication of certain types of aquatic vegetation, and the colonization of bodies of water with top-minnows that devour mosquito larvae, have been practiced.

The results seem no longer to be in doubt. Mosquito-breeding over an area of 200,000 acres has been virtually prevented.

This territory is now inhabited by more than 50,000 people. At the end of the Great War about one-third of the area of Italy was malarious and nearly sixteen million people were exposed to the disease. Between 1929 and 1933, 500,000 acres of land were freed from endemic malaria, and since then each year has witnessed the addition of thousands of acres of waste land to that already salvaged. The only other malaria campaign that rivals this in grandeur is the reclamation of the Panama Canal Zone.

It is easy to brag about such conspicuously successful campaigns for the eradication of malaria, but let us not forget that these are merely isolated instances. They demonstrate what can be done, but malariologists agree that malaria throughout the World is more widespread and more intense now than it was before the War in 1914, and that we have actually been forced back by our insidious foe.

METHODS OF MALARIA CONTROL

The fundamental methods of controlling malaria have already been stated in connection with the campaigns in Panama and in Italy. These projects were carried out under very favorable conditions and with an ample supply of funds. In many malarious regions, however, the inhabitants are financially unable to support satisfactory control measures, either because they are too poor or because there are too few of them for the size of the area that must be treated. Malaria can exist only where there are susceptible anopheline mosquitoes, and human beings who carry infective gametocytes in their blood and are accessible to the mosquitoes. It is therefore possible to control malaria in any region (1) by eradicating susceptible mosquitoes, or (2) by preventing susceptible mosquitoes from biting human beings who carry gametocytes, or (3) by treating these gametocyte-carriers so as to eliminate the parasites from their blood.

Many attempts have been made to sterilize gametocyte-carriers by treatment with quinine and with plasmoquine and atabrine, which are drugs that have recently been developed by the Germans by means of experiments on bird malaria, but with very little success. Protecting gametocyte-carriers from mosquitoes has long been recognized as an effective method of control when combined with other methods. Mosquitoes can be kept away from infected persons by screening houses, by the use of bed nets, by means of chemical substances that are repellant to the insects, or by stabling cattle, horses, or some other domestic animal nearby for the mosquitoes to bite in preference to human beings. One enthusi-

FIG. 124. Mark Twain got under his bed net where he stayed until all the mosquitoes were under the net with him; then he hopped out and slept on the floor.

astic malariologist advocates "a pig under every bed" as a protective measure since certain mosquitoes prefer pigs to man. Some of you no doubt are familiar with Mark Twain's method of using a bed net. He was never able to arrange his net so as to prevent the mosquitoes from getting in, so he waited until all the mosquitoes got under the net and then hopped out and slept on the floor.

DESTROYING MOSQUITOES

But the most effective method of control is the destruction of malaria-transmitting mosquitoes. At first all anopheline mosquitoes were supposed to be transmitting agents of human malaria. Later it was discovered that only a few species could be infected in the laboratory and not all of these became infected in nature, largely because they did not ordinarily bite human beings. Finally our malaria detectives have demonstrated that certain races within a single species of mosquito may be susceptible to malaria and other races resistant. Thus the common malaria mosquito of Europe, *Anopheles maculipennis,* has been separated into several susceptible races and several that are not; and these races can be distinguished by differences in the structure of their eggs.

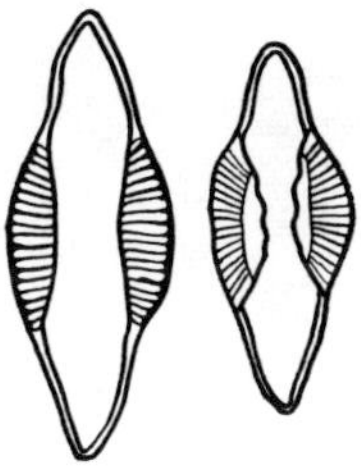

Eggs of mosquitoes showing differences in size and structure. *Anopheles maculipennis,* left *A. quadrimaculatus,* right.

This racial dissimilarity in susceptibility explains what has long been a mystery, that is, the absence of malaria from certain regions where large numbers of apparently susceptible anophelines were present. Now we know that these anophelines belong to non-susceptible races. The evidence is quite clear, therefore, that to control malaria by destroying mosquitoes, we must first determine for every area to be treated, which species or races are transmitting the parasites. Then we must learn where they breed and under what conditions, to what extent they feed on animals and human beings, where they hide during the daytime, how far they are accustomed to fly from their breeding places, and many other facts regarding their daily activities.

The destruction of adult mosquitoes is advocated as a control measure but is one of lesser importance. The insects often hide in some dark corner or under the house during the day and can be found and killed. Crawling about in dusty corners and under beds is neither a dignified nor pleasant occupation, especially when the weather is hot and muggy and the house is not particularly clean. No wonder the natives witness such a spectacle with amazement. As one Moslem boy was heard to remark, "Even if I were a Christian I would not do that" (Hackett). Mosquitoes may be trapped in various ways or killed by poison bait. Bats and dragon flies are well-known enemies that can be encouraged.

DESTROYING MOSQUITO LARVAE

Killing mosquito larvae is a much more effective method of control. The water in which the larvae live may be covered with oil or dusted with Paris green by a hand spray, or, in certain regions, like a smoke screen from an airplane. Natural enemies may be introduced into their breeding places, especially topminnows which feed largely on mosquito larvae and are never weary of well-doing.

The most gratifying anti-larval measure, however, is that of destroying breeding places. Ponds and pools may be drained; vegetation amid which larvae find protection can be cleared

FIG. 125. As the malariologist hunts for mosquitoes under the bed the little Moslem boy says, "Even if I were a Christian I wouldn't do that."

out of standing water and streams; depressions in which water collects can be filled; the surface of lakes and reservoirs may be lowered and raised again from time to time so as to expose the edges where larvae live; and in certain regions bodies of fresh water can be made salty by allowing sea water to flow in, and thus unfit for certain types of larvae.

As already noted malaria has been practically eradicated from certain regions, such as the Upper Mississippi Valley, the Panama Canal Zone and the Pontine Marshes, but it still holds sway over enormous areas on the earth's surface. Perhaps the most discouraging feature concerned with malaria control is the fact that every area differs in its requirements from every other area and demands special study before adequate control measures can be applied with hope of success. But what has been done can be done, hence there is no justification for pessimism and inactivity. "The important fact is not that there are unsolved, and at present insoluble, problems of malaria in the world. It is that there are thousands of

FIG. 126. We may eradicate malaria from a region but vigilant health officers are needed to keep it from returning.

communities oppressed for ages by the disease that could be liberated with a little thought, energy, and money which would not be difficult nowadays to assemble and apply. The mechanism of malaria transmission is so complicated and delicate that it never has been able to resist any long-continued sabotage. Persistence is more important than perfection, and whether control is a partial failure or a partial success depends on the point of view" (Hackett).

The necessity of persistence is well known to all malariologists, since malaria has a habit of returning if control measures are not maintained year after year. The school boy who said Milton was a great parasitologist because he wrote Parasites Lost and Parasites Regained got his facts a bit mixed but he undoubtedly referred to a condition that only too often follows a campaign for malaria control.

THERAPEUTIC MALARIA

As already noted the protozoa that live in man apparently have adopted the slogan "live and let live," since they exhibit little or no antagonism to one another. The effects of their presence on human parasites other than protozoa may, however, be quite pronounced, and may even transform such a terrible enemy as the malaria organism into a beneficent ally. Who would consider for a moment the deliberate infection of human beings with malaria? Boiling in oil should be the mildest punishment meeted out to such a monster. Yet thousands of patients in hospitals all over the world are now being treated with malaria without any protests whatever. This remarkable situation is due to the fact that when malaria parasites are deliberately inoculated into a person they can be eliminated again quite easily and the infection cured. If such a person were suffering from general paralysis, the malaria might actually relieve him of some of his symptoms.

General paralysis, which is known also as paresis, is a syphilitic disease due to a minute spiral organism that belongs to a group of micro-mites called spirochetes; this particular spirochete was baptized with the name *Treponema pallida.* It is a living corkscrew, and a homeless waif, sometimes allowed by scientists to associate with the bacteria and sometimes with the protozoa, but is usually left out in the cold altogether. We don't entirely approve of allowing malaria parasites to play with such loathsome creatures, but if the spirochetes are transformed into "good Indians" by the association we are willing to withdraw our objections. Ordinarily patients suffering from general paralysis die in from one to three years, although remissions may occur during which physical and mental conditions appear to be quite normal.

Medical literature contains many references to improvement

in psychopathic patients following acute attacks of malaria. Among the physicians who have referred to this phenomen were Hippocrates, Galen, Boerhaave, Sydenham, and more recently, Koster (1848), and Wagner-Jauregg of Vienna (1887). But the eyes of the medical profession were not opened to the efficacy of malaria-treatment of general paralysis until 1917, when the same Wagner-Jauregg inoculated nine patients and obtained three complete remissions. Since then this method of treatment for general paralysis of the insane and other grave nervous disorders has spread throughout the world and has come to be regarded as a routine procedure.

Patients are usually inoculated in the skin (intradermally) or in a muscle (intramuscularly) with one or two cubic centimeters of blood from another patient suffering from benign tertian malaria. Blood containing malaria organisms may be kept for ten days in a thermos bottle at the freezing point and in this condition sent long distances by mail. After eight or ten paroxysms at intervals of three days the patient is treated with quinine. About 30 per cent of general paralytics are cured by this treatment and a further 20 per cent greatly improved both physically and mentally.

The importance attributed by medical men at the present time to malaria and other fever-producing agents is indicated by the fact that on March 29, 30 and 31 of this year (1937) the First International Conference on Fever Therapy was held in New York City. More than two hundred eminent physicians, pioneers in this field, came from twenty-eight countries to participate in the conference. Fourteen countries sent officially appointed representatives of their ministries of health. In addition to the official delegates from the United States Army, Navy and Public Health Service, this country was represented by more than sixty leading physicians from twenty States.

Not the least important feature of malaria-therapy is the accumulation of data regarding the activities of human malaria parasites in man under controlled conditions. These data have helped us to devise better methods of protecting individuals who live in malarious regions from acute attacks, and to develop more satisfactory methods of treatment. We may be accused of being too enthusiastic when we predict that the terrors of malaria will be practically eliminated within the next two decades but certainly hope that this optimistic prophecy will be fulfilled.

I have attempted in this book to demonstrate how much more marvelous protozoa are than anything Alice encountered in her looking-glass dream. If now you don't vote in favor of

Fig. 127. Like the Hydra that Hercules conquered, when one malaria problem is solved, two spring up to take its place.

the protozoa I shall consider myself a very poor advocate. Anyway it has been a real pleasure to present some of the facts about these fascinating little creatures that are so important in relation to human welfare. But like the Hydra that Hercules conquered in Argos, which grew two heads for every one cut off, the study of one problem usually uncovers several others that were previously unsuspected, and hence we seem to be confronted by a greater area of unexplored territory than ever before. However, there is on need for us to feel downhearted since, as Robert Louis Stevenson remarked, "To travel hopefully is a better thing than to arrive, and the true success is to labor."

Conquistadores

Our concluding poem summarizes the history of the conquest of malaria and emphasizes the international character of the fight for the solution of this world problem. It is entitled, I believe appropriately, Conquistadores.

Both the Greeks and Ancient Romans
Bore the yoke of chills and fever,
Likewise through the Middle Ages.
Not until the savage Incas
Gave to us cinchona-tree bark
Did mankind do aught about it.
Static then the dread miasma
'Till in hot and dusty Algiers,
Laveran, the Frankish savant,
Spied the guilty little devil
Hiding in its red corpuscle;
Proved the germ of paludisme.
Followed Ross the bulldog Briton,
G-man of the Indian service;
Trapped the deadly female 'squito,
Public enemy of the birdies;
Paved the way for Roman Grassi
And anopheline detection.
Quickly then our General Gorgas
Freed the tropics of malaria;
Showed that research scientific
Must preceed the application;
Cleared Havana and the Isthmus;
Made the uniform immortal.
Nor must we forget the Germans
With their Chemische Fabriken,
Nor the innocent canaries,
Guinea-pigs with golden winglets,
Sacrificed for sake of science,
Test-tubes for teutonic genius,—
Plasmoquine and atebrine
Now a monument to their glory.
Thus we conquer nature's scourges;
Hasten on the goal,—millenium;
Frenchman, Briton, and Italian,
German and Americano,
Internationally minded;
Deathless these Conquistadores.

Appendix A

GENERAL REMARKS ABOUT PROTOZOA

The scientific study of animals is known as zoology and those who engage in this study are zoologists.

No one knows how many different kinds of animals are now in existence; about 840,000 have been described by zoologists. Fortunately for us, although every one of these differs from every other sufficiently to be recognized as a distinct species they possess various characteristics in common and can be arranged into groups. The principal groups are called phyla (singular, phylum). These phyla are usually arranged in a series according to their degree of complexity beginning with the simplest and ending with the most highly developed.

The phyla usually recognized by zoologists today may be listed as follows; the number after each scientific name indicates the approximate number of species that have been described in the phylum.

1. Protozoa (15,000 species).—One-celled animals.
2. Porifera (3000 species).—Sponges.
3. Coelenterata (5000 species).—Hydroids, polyps, jellyfishes, sea-anemones and corals.
4. Ctenophora (100 species).—Sea walnuts and comb jellies.
5. Platyhelminthes (6500 species).—Tapeworms, liver flukes and other flatworms.
6. Nemathelminthes (3500 species).—Thread worms or round worms.
7. Echinodermata (5000 species).—Starfishes, brittle stars, sea urchins, sea cucumbers and sea lillies.
8. Annelida (5000 species).—Segmented worms, such as earthworms and leeches.
9. Mollusca (78,000 species).—Snails, slugs, clams, oysters, devil fishes, etc.
10. Arthropoda (675,000 species).—Lobsters, crabs, centipedes, insects, spiders, ticks, etc.
11. Chordata (40,000 species).—Mostly vertebrates, such as lamprey-eels, sharks and rays, fishes, amphibians (frogs, toads and salamanders), reptiles (alligators, lizards, snakes and turtles), birds, and mammals or four-footed beasts.

The Protozoa are considered to be the simplest of all animals and hence

General Remarks about Protozoa

constitute the first phylum in the animal kingdom. Although most of them consist of but a single cell they differ widely one from another in structure, physiology, methods of reproduction and life cycles. They vary in size from those smaller than a red blood corpuscle, such as the malaria parasite, to relatively huge species, such as the gregarine that lives in the lobster, a species that may be two-thirds of an inch long. Most of the protozoa are invisible to the naked eye.

In spite of the small size and vast numbers of species of Protozoa it is not difficult to arrange them into classes, orders, families, genera, and species. The most useful characteristic is the nature of the locomotor organs if these are present.

Class I. Sarcodina (Gr. *sarx*, flesh).—Type: *Amoeba*. Protozoa that move by means of false feet, or pseudopodia.

Class II. Mastigophora (Gr. *mastix*, whip; *phero*, bear).—Type: *Euglena*. Protozoa that move by means of flagella.

Class III. Sporozoa (Gr. *spora*, seed; *zoon*, animal).—Type: *Monocystis*. Protozoa without motile organs but with a spore stage in their life cycle.

Class IV. Infusoria (Lat. *infusus*, crowded in).—Type: *Paramecium*. Protozoa that move by means of cilia.

The shapes of protozoa are extremely varied. Ameba has no fixed shape; many species are globular with radiating projections (Radiolaria); Euglena is spindle-shaped; Paramecium resembles a slipper; Vorticella a bell; Stentor a trumpet; some have definite dorsal and ventral surfaces; in fact, almost every conceivable shape seems to occur in this group.

Most of the Protozoa are either faintly colored or entirely without pigment. When coloring-matter is present it often consists of chlorophyll, or some allied substance, which is contained in chromatophores, e.g. Euglena. Drinking water is often colored red by *Euglena sanguinea*, or yellow by Uroglena; the surface of the sea is sometimes colored orange by vast numbers of Noctiluca, or red by Peridinium.

The simplest kind of locomotor organs are pseudopodia like those of Ameba. The pseudopodia of some species have a firm axial rod, and those of others may branch and fuse with one another. Flagella may be likened to very thin pseudopodia that have become permanent. They seem to be composed of long fibrils that are spirally wound. Cilia are smaller and more numerous than flagella; often they are fused together in groups forming large cirri, or side by side, forming membranelles as in the gullet of Paramecium.

An external covering may be absent from the body of Protozoa (Ameba) or may be present as a distinct cuticle (Paramecium). Shells may also occur; these consist of material secreted by the animal, e.g. chitin by Arcella, calcium carbonate by Globigerina, and silicon by Radiolaria, or are made up of foreign particles such as grains of sand (Difflugia).

Within the body of Protozoa are embedded one or more nuclei, vacuoles of several kinds, and frequently plastids. A nucleus is always present. Some Protozoa have two kinds of nuclei (e.g. Paramecium), a macronucleus and a micronucleus. During binary division the chromatin of the nucleus may form distinct chromosomes. Vacuoles are of several kinds such as permanent globules of liquid, contractile vacuoles and food vacuoles.

Many Protozoa possess plastids; these are usually bodies of starchy food material, or colored bodies called chromatophores, such as occur in Euglena. Besides these, many other substances may be present, such as food material, indigestible matter, oil drops, grains of sand, etc.

The food of Protozoa consists of organic matter both vegetable and animal. Bacteria, diatoms, and other Protozoa form a large part of the bill of fare. Such forms as Euglena do not ingest solid food, but manufacture it by means of chlorophyll.

Usually some structure is present which aids in the ingestion of food, but in the Sarcodina, like Ameba, there is no mouth, and food is engulfed at any point on the surface. The flagella of many flagellates and the cilia of ciliates draw or drive food particles toward the mouth and down into the gullet, at the end of which a food vacuole is formed. The Suctoria capture their prey with their tentacles and suck the contents into the body. Many parasitic Protozoa take food directly through the surface of the body.

Digestion takes place in the food vacuoles. The surrounding protoplasm secretes ferments which enter the vacuoles and dissolve certain food substances. Undigested matter is cast out at any point (Ameba), or at a particular spot (Paramecium), or through a definite anal opening (Balantidium). Digested food passes out into the cytoplasm and is assimilated, i.e. is transformed into protoplasm. Oxygen is necessary before life activities can be carried on, and carbon dioxide is given off. This is respiration. The oxygen is taken in through the body-wall. It combines with protoplasm, i.e. oxidation takes place. Carbon dioxide and other waste matter in solution are by-products. These by-products pass out through the body-wall, and probably by way of the contractile vacuole. The contractile vacuole may therefore be called a primitive excretory organ.

The Protozoa thus carry on many of the activities characteristic of the higher organisms without the aid of the systems of organs we usually associate with these functions.

Protozoa move from place to place either by creeping over the surface of objects (Ameba) or by free swimming. The locomotor organs are pseudopodia, flagella, and cilia. In some Protozoa fibrils (myonemes) are present just beneath the cuticle (Stentor, Vorticella); these are capable of contraction and can change the shape of the animal. In the stalk of Vorticella the fibrils are agents for moving the bell from place to place.

Protozoa are capable of spontaneous activities and respond to a number of different external stimuli, which are changes in their environment. The study of the behavior of the lower organisms has become quite prominent since the year 1900 and has led a prominent investigator in this field to the following conclusion. "All together, there is no evidence of the existence of differences of fundamental character between the behavior of the Protozoa and that of the lower Metazoa (many-celled animals). The study of behavior lends no support to the view that the life activities are of an essentially different character in the Protozoa and the Metazoa. The behavior of the Protozoa appears to be no more and no less machine-like than that of the Metazoa; similar principles govern both." (Jennings, Behavior of the Lower Organisms, p. 263.)

Reproduction.—The usual method of reproduction in the Protozoa is that of binary division. This occurs in such forms as Ameba, Euglena, and Paramecium. During binary division the body of the protozoon divides into two approximately equal parts, the daughter-cells. When the division of the protozoon is unequal, the process is spoken of as budding or gemmation. A third method of reproduction is by the formation of spores as in the Sporozoa.

Appendix B

GLOSSARY

Every science has its own vocabulary and zoology and protozoology are no exceptions. Although scientific terms have been reduced to a minimum in this book, it was found necessary to use a number of words that might not be familiar to general readers, hence as many of these as seemed practicable are listed here and defined.

Absorption. The taking in of fluid material by a living cell.

Adaptation. An advantageous variation enabling an organism to live successfully in its environment.

Aerial. Inhabiting or frequenting the air.

Amoeboid Movement. A type of movement resembling that of an Amoeba.

Anus. The opening at the posterior end of the alimentary canal.

Appendix. A thick-walled, finger-like process at the end of the cecum in certain mammals.

Arboreal. Inhabiting or frequenting trees.

Arteries. Tubes that convey blood from the heart to all parts of the body.

Asexual Reproduction. Reproduction without sex cells.

Assimilation. Changing food into living substance by an organism.

Behavior. All of the reactions of an animal to its environment.

Bilateral. Having the sides symmetrical.

Binary Fission. Division of a cell into two daughter cells.

Biogenesis. The doctrine that all living things arise from preexisting living things.

Biology. The science of life and living things.

Buccal. Pertaining to the mouth; oral.

Caecum. An appendage of the intestine that ends blindly.

Carbohydrate. An organic compound composed of carbon, hydrogen, and oxygen and with two atoms of hydrogen to one of oxygen; examples, starches, sugars, cellulose.

Carnivorous. Feeding on animal food.

Cell. The unit of structure in organisms. A mass of protoplasm containing a nucleus or nuclear material.

Chlorophyll. The green coloring matter in plants and certain animals active in photosynthesis.

Chromatin. A substance in the nucleus that becomes deeply stained by certain dyes.

Chromatophore. A pigment-bearing cell or plastid.

Chromosome. One of the deeply staining bodies formed within the nucleus during mitosis.

Cilium. A minute hair-like cytoplasmic process projecting from a cell or organ.

Class. A main subdivision of a phylum.

Coagulation. The change from a fluid to a thickened, congealed condition, as in the clotting of blood.

Cold-blooded. Having blood of about the temperature of the surrounding air or water; fishes, amphibians, and reptiles are cold-blooded vertebrates.

Colony. An association of a number of individuals of one species.

Commensalism. A close association of two species of animals neither of which injures the other; eating together at the same table.

Conjugation. The temporary union of two Protozoa during which chromatin material is interchanged.

Contractile Vacuole. A reservoir in Protozoa into which flow water and waste products of metabolism which are expelled to the outside.

Cutaneous. Pertaining to the skin.

Cyst. An organism around which a resistant wall has formed.

Cytology. The science of cell structure.

Dermal. Pertaining to the skin or external covering of the body.

Digestion. The breaking down of food into simpler material that can be absorbed.

Diurnal. Pertaining to the period of daylight, as distinguished from the night.

Ectoparasite. A parasite that lives on the outside of the body of its host.

Endoparasite. A parasite that lives inside of the body of its host.

Epithelium. A type of cell layer that covers external and internal surfaces.

Evolution. The doctrine that species of animals have evolved by change from preexisting species.

Excretion. The discharge of waste products of metabolism; also the substance discharged.

Faeces. The unabsorbed residue of digestion; excrement.

Fauna. The total of the animal life of a given region or period.

Fertilization. The fusion of two germ cells, egg and spermatozoon, to form a zygote.

Fission. The division of a cell or organism into two or more parts; a type of asexual reproduction.

Fly. A species of insect belonging to the order Diptera; used commonly for certain species belonging to other orders of insects.

Free-Living. Existence independent of other organisms; not parasitic.

Function. The special, normal, or proper action of any part or organ of the body.

Gamete. A germ cell capable, either alone or after fusion with another germ cell, of further development.

Gametocyte. A cell capable of producing one or more gametes.

Gene. The unit of inheritance. Something in a chromosome responsible for the development of a character. A factor or determiner.

Genital. Pertaining to the reproductive organs.

Genus. The principal subdivision of a family.

Germ Cell. A cell which alone, or in combination with another germ cell, is capable of reproduction.

Germ-plasm. The material basis of heredity; the chromatin.

Gill. The breathing organ of an animal that lives in the water.

Gland. One or many associated cells that elaborate and discharge secretions.

Habitat. The environment in which an organism normally lives.

Haemoglobin. The red substance in the blood that carries oxygen.

Herbivorous. Feeding on vegetation.

Heredity. The transmission of genes and the characters for which they are responsible from parents to offspring.

Holophytic. Plant-like nutrition involving photosynthesis.

Holozoic. Animal-like nutrition involving the ingestion and digestion of organic material.

Host. An organism within or upon which a parasite lives.

Hypothesis. A proposition that is assumed to be true and can be used in studying related phenomena.

Ingestion. The act of taking in, especially food.

Invertebrate. An animal without a spinal column or backbone.

Irritability. The property of responding to stimuli.

Larva. The young of various animals that usually differ in appearance from the adult.

Lumen. The cavity of a hollow, tubular organ, such as intestine, gland, duct or blood vessel.

Lymph. Plasma and white blood corpuscles that bathe the tissue cells.

Maggot. An insect larva, especially the footless larva of a fly.

Migration. Movement of organisms (usually in groups) from one region to another.

Morphology. The science of form.

Myoneme. A type of contractile fibril in certain Protozoa.

Natural Selection. The "selection" in nature of those individuals of species best fitted to survive in a particular environment.

Nocturnal. Active by night.

Nucleus. A usually spherical body within the cell that contains the chromatin.

Oocyte. The ovarian egg before maturation.

Oral. Pertaining to the mouth or ingestive opening.

Order. A main subdivision of a class.

Organ. A combination of tissues for the performance of a particular function.

Organism. A member of the animal or vegetable kingdom.

Ovum. The female gamete or egg.

Parasite. An organism that lives on or within and at the expense of another organism.

Parthenogenesis. Development of an unfertilized egg.

Pathogenic. Giving rise to disease.

Peristalsis. The forcing of food along the intestine by rhythmical contractions of the intestinal wall.

Peristome. The region surrounding the mouth cavity.

Phosphorescent. Shining with a faint light or luminosity like that of phosphorus.

Photosynthesis. The formation of complex compounds by chlorophyll in the presence of light.

Phylum. A main division of the animal kingdom.

Physiology. The study of function.

Pigment. A substance in the tissue that gives color to the body.

Plasma. The liquid part of the blood or lymph.

Predaceous. Capturing living animals for food.

Prehensile. Fitted for grasping or holding.

Proteins. The principal constituents of protoplasm consisting of complex molecules containing nitrogen.

Protoplasm. The living substance of which plants and animals are composed.

Pure Line. A group of individuals of one species that contain the same genes and are descended from a homozygous ancestor.

Rectal. Pertaining to or connected with the rectum, the terminal portion of the intestine.

Reproduction. The production by an organism of others of its kind.

Respiration. Internal respiration is the exchange of carbon dioxide for oxygen by protoplasm. External respiration is the entrance of oxygen into the body and exit of carbon dioxide from the body.

Response. The reaction of an organism to a stimulus.

Scavenger. An animal that feeds on dead or decaying animal matter.

Secretion. The production by the protoplasm of a substance of use to the organism; also the substance produced.

Soma. The body tissue in contrast to the germinal tissue.

Special Creation. The theory that each species of animal was specially created and did not arise by the process of organic evolution.

Species. A group of individuals that resemble one another closely, and whose characters are pronounced and not obliterated by intermediate forms.

Spontaneous Generation. The theory that organisms arise directly from inorganic substances.

Sporulation. Formation of or conversion into spores.

Stimulus. A change within an animal or in its environment that brings about a response.

Symbiosis. Living together of two species of organisms each of which is necessary to the other.

System. An assemblage of parts or organs of the same or similar tissues. A group of organs that together perform some function.

Tissue. An aggregation of similar cells.

Trophozoite. The active, vegetation stage in the life-cycle of a protozoon.

Urea. A nitrogenous substance resulting from animal metabolism.

Vacuole. A minute vesicle in the cytoplasm of a protozoon.

Vegetative. Pertaining to nutrition.

Veins. Blood vessels that convey blood to the heart. Tubes that form the framework of an insect's wing.

Vertebrate. An animal possessing a backbone consisting of vertebrae.

Visceral. Pertaining to the interior organs of the body, especially those within the abdominal cavity.

Warm-blooded. With body temperature at practically a constant temperature regardless of the temperature of the environment; this occurs in birds and mammals.

Appendix C

LITERATURE ON PROTOZOA

The literature of Protozoology is contained principally in textbooks and reference books on the subject, textbooks and reference books on tropical medicine, and in journals of parasitology, tropical medicine, zoölogy and general biology. The author lists below what he believes to be the more important of the comprehensive books and monographs in English and also a list of the journals in English that include, from time to time, papers on protozoology.

TEXTBOOKS AND REFERENCE BOOKS

Becker (E. R.). 1934. Coccidia and Coccidiosis. 147 pp. Ames.

Blacklock (D. B.) and Southwell (T.). 1931. A Guide to Human Parasitology. 271 pp. London.

Boyd (M. F.). 1930. An Introduction to Malariology. 437 pp. Cambridge, Mass.

Byam (W.) and Archibald (R. G.) (Editors). 1921–1923. The Practice of Medicine in the Tropics. Vol. 1, 1921, 856 pp. Vol. 2, 1922, pp. 857–1683. Vol. 3, 1923, pp. 1685–2550. London.

Calkins (G. N.). 1901. The Protozoa. 347 pp. New York.

Calkins (G. N.). 1933. The Biology of the Protozoa. 2nd ed. 607 pp. New York.

Craig (C. F.). 1926. Parasitic Protozoa of Man. 569 pp. Philadelphia.

Craig (C. F.). 1934. Amebiasis and Amebic Dysentery. 315 pp. Baltimore.

Dobell (C.). 1919. The Amoebae Living in Man: a Zoölogical Monograph, 155 pp. London.

Dobell (C.) and O'Connor (F. W.). 1921. The Intestinal Protozoa of Man. 211 pp. London.

Edwards (F. W.). 1932. Genera Insectorum. Diptera. Fam. Culicidae. 258 pp. Tervueren.

Hackett (L.). 1937. Malaria in Europe. 336 pp. Oxford.

Hall (M.). 1936. Control of Animal Parasites. 162 pp. Evanston.

Hegner (R. W.) and Taliaferro (W. H.). 1924. Human Protozoölogy. 597 pp. New York.

Hegner (Robert). 1927. Host-Parasite Relations between Man and His Intestinal Protozoa. 231 pp. New York.
Hegner (Robert) and Andrews (Justin). 1930. Problems and Methods of Research in Protozoölogy. 532 pp. New York.
Knowles (Robert). 1928. An Introduction to Medical Protozoölogy. 887 pp. Calcutta.
Kudo (R. R.). 1931. Handbook of Protozoölogy. 451 pp. Baltimore.
Manson-Bahr (P. H.). 1936. Manson's Tropical Diseases. 10th ed. 1003 pp. London.
Matheson (Robert). 1929. A Handbook of the Mosquitoes of North America. 268 pp. Baltimore.
Minchin (E. A.). 1912. Introduction to the Study of the Protozoa. 520 pp. London.
Ross (Sir Ronald). 1911. The Prevention of Malaria. 2d ed. 711 pp. London.
Rudolf (G. de M.). 1927. Therapeutic Malaria. 223 pp. Oxford.
Smith (Theobald). 1934. Parasitism and Disease. 196 pp. Princeton.
Taliaferro (W. H.). 1929. The Immunology of Parasitic Infections. 414 pp. New York.
Thomson (J. G.) and Robertson (A.). 1929. Protozoölogy. 376 pp. London.
Wenyon (C. M.). 1926. Protozoölogy, 2 vols. 1563 pp. London.

JOURNALS THAT CONTAIN CONTRIBUTIONS ON PROTOZOA

American Journal of Hygiene. Baltimore. Vol. I. 1921.
American Journal of Tropical Medicine. Baltimore. Vol. I. 1921.
Annals of Tropical Medicine and Parasitology. Liverpool. Vol. I. 1907.
Indian Journal of Medical Research. Calcutta. Vol. I. 1913.
Journal of Parasitology. Urbana. Vol. I. 1914.
Journal of Tropical Medicine and Hygiene. London. Vol. I. 1898.
Parasitology. Cambridge. Vol. I. 1908.
Tropical Diseases Bulletin. London. Vol. I. 1912.
Tropical Veterinary Bulletin. London. Vol. I. 1912.

ACKNOWLEDGMENTS

This book grew out of a series of six lectures, known as the Messenger Lectures, delivered before a general audience at Cornell University, Ithaca, New York, in April, 1937. The author is indebted to the Cornell University Press for waiving their rights of publication.

The cartoons were made under the author's direction by Miss Dagmar Haugen (Chapters I, II, IV, VI, VIII) and Miss Betsey Garrett (Chapters III, V, VII, IX, X).

Acknowledgments

Most of the other illustrations were drawn by the author. A few of them were selected from periodical literature on protozoology. The Macmillan Co. kindly permitted the use of the drawings of the parasites of man, life cycle of malaria parasite, ameba and relatives, paramecium and relatives, trypanosome, leishmania, and euglena. The photographs of the dysentery ameba, balantidium in tissues and malaria in mosquito were obtained from the Army Medical Museum. Boyd furnished the photographs of malaria in the blood; Herms and Snyder, the termite drawings; and Lineburg the drawing of volvox.

INDEX

All numbers refer to pages. Words in *italics* are genera or species. Numbers in **bold face** type refer to pages on which there are figures.

Abildgaard, 244
Acanthocheilonema, 197
Aciduric bacteria, **178**
Adult, mosquito, **247**
Ague, 239
Algae, **57**
Amebas, contractile vacuole, 129; cyst, **129**, 132, 161; distribution, 39; dysentery, **129**, 164; large intestine, 129; life cycle, 129, 131; monkeys, 148; nuclei, **131**; oral, **120**, 147; parasitized by *Sphaerita*, **62**; size, 10; soil, 5; whip-bearing, **47**
Amebiasis, 167; epidemics, 170; treatment, 168
Amebic abcesses, 168
Amebic ulcer, **166**
Amoeba proteus, **3**, **33**; behavior, 28, 37; division, 38; food habits, 34; food selection, 37; life of, 32; physiology, 36; relatives of, 39; sex, 38; size, 32; structure, 34
Amphileptus branchiarum, **23**, 24
Anaerobes, 179
Anayodin, 169
Anemia, malaria, 239
Animal diet, Eskimos, 103
Animal mind, 27
Anopheline, mosquitoes, 245, **247**, **249**, 259
Antelope, 200
Anthropology, 136
Antimony, 208
Arcella, **21**, **22**, 40
Aricayo, 90
Atebrine, 258
Atkins, 196
Avoiding reaction, *Euglena*, 43; *Paramecium*, **28**
Axostyle, 132

Bacillus malariae, 242
Bacteria, intestinal, **178**
Bagdad sore, 209
Bailly, 240
Balantidium, 51; dysentery, 134; Filipino pigs, 105; man, 133; monkey, 138, 152; pig, 138; starch diet, **180**; structure, 133
Balantidium wenrichi, 152
Balboa, 251
Bardsley, 168
Bastianelli, 248
Bass, 232, 236
Bats, protozoa, 105
Baugio, 103
Bayer 205, 201
Bedbug, 211
Behavior, of ameba, 37
Bence-Jones, 240
Bentley, 207
Bignami, 248
Bilateral symmetry, 124
Bladderwort, 111, **112**
Blepharisma lateritia, **23**, 24
Blind alley theory, 135
Blood-inhabiting protozoa, 73, 78; bats, 105; distribution, 99; leishmanias, 205; malaria, 215; man, **134**; transmission, 79; trypanosomes, **189**
Bohol, 107
Boil disease, 69
Bolestra, 241
Bontoc Igorots, 103
Brown, 202
Bruce, 79, 198, 199
Bugs, vectors, 65

Cabatuan, 89
Cagayan valley, 87
Canaries, malaria, 227

Carbarsone, 169
Carnivorous animals, 174
Carnivorous diet, 175
Casteliani, 198
Cecum, chick, 109; monkey, **147**
Cells, division of labor, 46
Cells, body, 46; germ, 46; somatic, 46
Chagas, 205
Chagas disease, 202; California, 203; history, 205; reservoir hosts, 204; symptoms, 204; transportation, 213; transmission, 204; treatment, 205
Chicago, amebiasis, 170
Chicken coccidia, 78
Chicks, living test-tubes, 109
Chills and fever, 220
Chilomastix mesnili, 132; cyst, **133**
Chilomonas, 35
Chimpanzee, 154
China, 104
Chiniofon, 169
Chepigana, 102
Chiquimula, 185
Chlorophyll, 41
Choanoflagellates, **47**
Christy, 198
Chu, 87
Chucumique river, 103
Cilia, 48
Ciliates, 48; balantidial, 105; cockroach, 244; earthworm, 67; horses and cattle, 74, **75**; *Hydra*, 65; Opalinas of frogs, 74; parasitic, **23**; parasitized by Suctoria, **63**; soil, 5
Cinchona bark, malaria, 251
Cirri, 51
Clark, 196
Clone, 21, 27
Coccidia, chickens, 78; feline, 127; human, **126**; life cycle, 127; rabbit, 15, **78**; rats, 140; species, 126; transmission, 126
Colon, horses, **75**
Columbus, 251
Commensalism, 25
Conjugation, **23**
Contractile vacuole, 37, 129
Cosmozoa Theory, 18
Crescent, **219**, 221
Crocodiles, 87, 92
Cruz, 205
Culex fatigans, 246; *nemerosus*, 248; *pipiens*, 225, 245, **247**
Culture medium, 109
Cunningham, 206
Curtice, 72
Cynocephalus volans, 107
Cyst, **55**, 74, 104, 122, **125**, **129**, 165; resistance, 161

Danguise, 196
Darien, Panama, 102
Darwin, 20, 240, 243
Death, natural, 46
Delhi boil, 206, 209
Diabetes, malaria, 232
Didinium, **49**
Dientamoeba fragilis, 130, **131**
Difflugia, **40**
Dinobryon, **4**
Diseases of wild animals, 81
Donovan, 206
Dourine, 81
Drosophila, 25
Drug resistance, 169
Dufour, 244
Dujardin, 15
Dum-dum fever, 206
Dutton, 79, 197, 198
Dysentery ameba, **129**, 164; amebic, 165

Earthworm, ciliates, **67**; sporozoa, 54, 55
Egg, malaria, **217**, 223; mosquito, **247**; *Volvox*, **44**, 45
Ehrenberg, 16
Eimeria, 126
Eimeria stiedae, 15
Eklund, 241
El Real, 102
Emetine, 168
Empedocles, 250
Encephalitis lethargica, 194
Endamoeba coli, 130, **131**; *histolytica*, **129**; *muris*, 139
Endolimax nana, 130, **131**
Endomixis, 229
Eskimos, 103

Espundia, 210
Euglena, 15; bladderwort, 113; characteristics, 41, **42**; eye spot, 42; movements, **41**
Euglenoid movements, **41**
Evans, 79
Exflagellation, **217**, 221, 226

Fertilization, 26, **217**, 222; *Volvox*, 45
Fever, malaria, 238
Filipinos, 99
Filterable virus, 71
Firth, 206
Fish parasites, 67
Flacherie, 71
Flagellate diarrhea, 177
Flagellates, 41; blood-inhabiting, 78; Chagas disease, 204; colonial, **5**, 43, **47**; drinking water, 3; large intestine, 132; leishmaniasis, **207**; monkeys, 147; oral, **123**, 147; parasitic, 76; plants, **64**, 108; sleeping sickness, **195**; small intestine, 124; soil, 5; symbiosis, 58
Flagellum, **189**, 190
Flea, trypanosomes, **193**
Fleas, jumping, 88
Fly, trypanosomes, 194
Flying lemur, 107
Food vacuole, 35
Food robber, 61
Forde, 197
Forest yaws, 210
Fourneau 309, 202
Frog, opalinas, **74**

Gametocyte, **217**, 220
General paralysis, 262
Genetics, 25
Geographical distribution, 120
Germ cells, *Volvox*, **44**
Germ plasma, 19
Giardia lamblia, 163; *muris*, 139
Giardias, 15; birds, 85; civet cat, **84**, 85; cyst, **125**; guinea-pig, 85; human, 123; mice, 84; monkey, 147; prevalence, 125; rabbit, **84**; species, 84; transmission, 84, 125
Globigerina, 12
Glossina, 194, **199**, 200
Goethe, 256
Gorgas, 253
Grasshopper, gregarine, **55**
Grassi, 248, 249
Greeks, malaria, 240, 250
Gregarinida, **55**
Gruby, 79, 80
Guatemala, 182
Guerin, 196

Hackett, 238, 261
Haemozoin, 219
Haiti, malaria, 251
Helcosoma tropicum, 207
Hematochrome, 42
Hereditary transmission, 71, 78
Heterakis, 77
Heterotrichida, 24
Hexamita, 156, 176
Hippocrates, malaria, 240
Histomonas, **77**
Holotrichida, 24
Homo, 136
Host-parasite specificity, 85
Hydra, trichodinas, **64**
Hydrogen-ion concentration, 178
Hyperparasitism, 62
Hypotrichida, 25

Ichthyopthirius multifiliis, **68**
Ilagan, 88
Intestinal protozoa, 73; carnivorous diet, 175; distribution, 99; life cycles, 163; monkeys, 147; new species, 110; protection from, 171; starch diet, 182; transmission, 160
Immunity, trypanosomes, 192
Incas, quinine, 251
India, malaria, 253
Infusoria, 48, 51, 267
Intermittent fever, 239
International highway, 212
Iodamoeba, 130, **131**
Ipecacuanha, 168
Irritability, 27, 37
Isospora hominis, **126**

Jamestown, malaria, 252
Johns, 232

Kala-azar, 205; geographical distribution, 205; history, 206; parasites, **207**; South America, 211; symptoms, 208; transmission, 208, 210; treatment, 208
Kalingas, 87
Karyosome, amebas, **131**
Kerona pediculus, **23**, 25
Kessel, 248
Kilborne, 72, 244
Kissing, 121
Kissing bug, 202, **204**
Klebs, 242
Kleine, 199
Koch, 139, 248
de Kruif, 72

Lafont, 64
La Palma, 102
Larva, mosquito, **247**
Laveran, 241
Leeuwenhoek, 13
Leidy, 244
Leidyopsis, **59**
Leishman, 206
Leishman-Donovan bodies, 207
Leishmania bodies, **203**, 205
Leishmania braziliensis, 207, *donovani*, **207**
Leishmaniasis, 205; America, 210; dermal, 207, 209; history, 206; transmission, 208, 210; visceral, 207
Leuckart, 244
Lewis, 79
Linnaeus, 32
Low, 198
Loxodes rostrum, **23**, 24
Luzon, 106

MacCallum, 226
Macrogamete, 128, **217**, 222
Macrogametocyte, **217**, 220
Macrophage, 206
Magac river, 89
Malaria, anemia, 239; aestivo-autumnal, 238; bad air, 239; birds, 224; carriers, 257; control, 254, 256; cultivation, 232, 235; diabetes, 232; discovery of parasite, 241; drinking water, 240; economic importance, 237; exflagellation, **217**, 221; fertilization, **217**, 222; general paralysis, 262; Gorgas, 253, 255; Greece, 250; history, 239; immunity, 228, 235; incubation period, 220; India, 253; latency, 229; life cycle, 215, **217**, 226; malignant tertian, 221, 239; merozoite, **217**, 220; monkeys, 138; mosquitoes, 245; New World, 251; oöcyst, **217**, 223; oökinete, **217**, 223; Panama, 253; parasite number, 221; paresis, 262; Paris green, 260; periodicity, 230; phagocytes, **217**, 220, 234; pigment, 219, 239; Pontine marshes, 255; problems, 228, 234; quartan, **221**, 239; relapse, 233; Rome, 250; schizogony, **217**, 220; schizont, **217**, 220; screening, 258; segmentation, **217**, 220; sexual stages, **217**, 220; soil, 240; spleen, 239; symptoms, 228, 239; temperature, 233; tertian, 221, 239; therapeutic, 262; toxins, 234; transmission, 227, 239, 244; trophozoite, **217**, 218, 219
Mal de caderas, 81
Man, protozoa, 118, **119**
Manipol, 87
Manson, 197, 198, 249
Mark Twain, 258
Marsh fever, 239
Mastigamoeba, **47**
Mastigina, **47**
Mastigophora, 41, 267
Mead, 240
Meckel, 242
Melanin, 219
Melnikoff, 244
Merozoite, 127, **217**, 220
Metazoa, **31**
Mice, sarcosporidia, 70
Microgamete, 128, **217**, 222, 244
Microgametocyte, 127, **217**, 220, 222, 244
Microsporidia, **71**
Milkweed flagellates, **64**
Monkeys, 135; *Dientamoeba*, 148; Philippine, 142; protozoa, 136, 145

Mosquito, adult, **247**, 260; bird malaria, 225, 245; destruction, 258; eggs, **247**, **259**; larva, **247**, 260; life cycle, **247**; malaria, 216, **217**, 245; pupa, **247**; salivary glands, 216, **217**, **223**; stomach, **217**, **223** strains, 259
Mouth, ameba, **120**, flagellate, **123**
Munro, 240
Muscle parasites, 70
Myxosporidia, 68

Negritos, 99, **101**
Nosema disease, 70
Nuclei, amebas, **131**
Nummulites, **11**
Nyctotherus ovalis, **23**, 24

Oldham, 240
Olongopo, 99
Oöcyst, **78**, **217**, **223**
Oökinete, **217**, 223, 248
Opalina, frogs, **74**, **173**; parasitized by amebas, **62**
Opercularia stenostoma, **23**, 24
Oriental sore, 206, **207**, immunity, 209; scientific name, 207; transmission, 210
Oscillaria, 243
Oxytricha fallax, 23, 25
Oxytrichidae, 2

Paludisme, 239
Paludismo, 239
Panama, 102, 149; malaria, 251, 253
Parabasal body, **189**
Paramecium, **48**, 229; anal aperture, 16; avoiding reaction, **28**; behavior, 28; bladderwort, 113; captured by ameba, **36**, conjugation, **26**, 49; immortality, 19, **20**; pitcher plant, 114; reproduction, 9; spiral locomotion, **48**; symbiosis, 57; trichocysts, **49**
Parasite, definition, 57
Parasites of protozoa, 61
Parasitism and evolution, 141
Paresis, 262
Paris green, 260
Parthenogonidium, **44**, 45
Pasteur, 18, 70
Pathogenic protozoa, 73
Pausanias, 250
Pearce, 202
Pébrine, 70
Pelletier, 168
Pericles, 250
Periodicity, malaria, 230
Peritrichida, 24
Peru, 251
Phagocytes, kala-azar, 208; malaria, 217, 220, 234
Philippine protozoa, 106
Photosynthesis, 41, 58
Phytomastigina, 46
Pig, balantidia, 105; scavengers, 103; malaria, 258
Pigment, malaria, 219, 239, 242, **243**
Pigmies, 99, **101**
Pioneering with protozoa, 115
Piroplasma donovani, 206
Pitcher plant, 114
Plagiotomidae, **23**, 24
Plasmodium, *brazilianum*, 139; *cathemerium*, 231; *cynomolgi*, 139; *falciparum*, 238; *inui*, 139; *knowlesi*, 139; *kochi*, 139; *malariae*, 238; *vivax*, 238
Plasmoquine, 257
Polar filament, **69**
Pontine marshes, 255
Prevalence, 120
Prevention, 120
Prophylaxis, 169
Protein diets, 178
Proteolytic bacteria, **178**
Proterospongia, **47**
Protista, 61
Protophyta, 62
Protoplasm, 16
Protozoa, behavior, 27, 269; bladderwort, 111, **112**; blood-inhabiting, 73; carnivores, 174; carnivorous diet, 175; classification, 31; coloration, 267; commensalism, 61; control, 162; cuticle, 267; digestion, 268; discovery of, 13; drinking water, 3; Earth's crust, 11; ectoparasites, 65; evolution, parasitic, 23; Filipino animals, 108; Filipino bats, 105; Filipino pigs, 105; fish parasites, 67; food, 268; free-living vs. para-

sitic, 23, 25; immortality of, 19; intestinal, 73; large intestine, 128; locomotor organs, 267; man, 118, **119**; monkeys, 136, 145; mouth, **120**, 160; movement, 268; muscle tissue, 70; number of species, 10; nucleus, 268; oral, **120**; organic evolution, 20; parasitic 61; physiology, 7; pioneering, 115; pitcher plant, 114; plastids, 268; prevention, 120; reproduction, 9, 269; respiration, 268; sea, **6**; sexual phenomena, 26; shapes, 267; shells, 267; silkworms, 70; small intestine, 124; soil, 5; spontaneous generation, 16; symbiosis, 57; tissue, 134; transmission, hereditary, 71; transmission, oral, 160; transmission, termite, 60; transmission, water, 171; types of association, 25
Pseudopodia, 32, **33**
Pupa, mosquito, **247**
Pure line, 21
Putrefactive bacteria, 178
Pyorrhea, 121

Quinine, 252

Rabbit, coccidia, **78**
Racial, resistance, 99, susceptibility, 99
Radiolaria, **13**
Rat, trypanosome, 191
Red blood corpuscles, malaria, 217
Redi, 17
Relapse, malaria, 233
Reservoir hosts, 81
Response, 37
Rhodesia, 199
Rio Chico Indians, **102**
Romans, malaria, 250
Rogers, 169, 207
Ross, 206, 220, 226, 244
Roundworm, 77

Salisbury, 241
Salmon, 72
Sandfly, **210**
Sarcocystin, 70
Sarcocystis, 134
Sarcode, discovery of, 15
Sarcodina, 40, 267
Sarcosporidia, **70**
Schizogony, **217**, 220
Schizont, **217**, 220
Segmentation, **217**, 220
Sex, in *Volvox*, 43, **44**
Sheep, sarcosporidia, 70
Sili, 91
Silkworm parasites, 70, **71**
Sleeping sickness, 194; control, 199; history, 196; theories, 197; treatment, 201
Smith, Theobald, 72, 244
Soil protozoa, 5
Special Creation, 18
Species, definition, 18
Spermatozoa, **44**, 45
Sphaerita, **61**
Spirochete, 262
Spleen, malaria, 239
Sponge, ancestor, 47
Spontaneous generation, 16
Spore-bearers, 54
Spores, **69**, **70**, **71**
Sporozoa, 54, 78, 267; fish, 68; spores, **69**
Sporozoa foruncuIosa, 206
Sporozoite, **126**, 216, **217** **223**, 224, 245
Sporulation, **217**, 231
Starch diet, 180
Stegomyia, 245
Stentor, **48**, **51**
Sternberg, 240
Stevenson, 264
Stimulus, 37
Stinging cells, *Hydra*, **66**
Stomach, cattle, 75
Streblomastix, 59
Struggle for existence, 81
Stylonichia, **51**
Suctoria, **52**, 53; parasitize ciliates, **63**
Sun animalcule, **40**
Surra, 81
Symbiosis, 57
Synura, 4
Syphilis, horse, 81
Syzygy, **55**

Tadpoles, opalinas, **74**; trichodinas, 66
Taliaferria clarki, 152

Tapeworm, 244
Tarsier, **107**
Tartar emetic, 208
Tentacles, of Suctoria, **52,** 54
Termites, **58**
Termopsis angusticollis, 58
Texas fever, 72
Therapeutic malaria, 262
Ticks, Texas fever, 73
Todd, 198
Torti, 239
Tracheliidae, 24
Transmission, termite protozoa, 60
Transportation and disease, 212
Treponema pallida, 262
Triatoma, 202, **204**
Trichocysts, **49**
Trichodina, 65, **66**
Trichomonas, 76; cattle, **77**; chimpanzee, 156; diet, effects of, 176; large intestine, 132; monkeys, 147; mouth, **123;** pigeon, 77; rats, 140; termites, 59; transmission, 164
Trichomonas hominis, 132
Trichonympha, 59
Troglodytella, **138,** 156
Trophozoite, **217,** 218, **219**
Tropical diseases, 212
Trypanosoma gambiense, 198; *lewisi*, 191
Trypanosome fever, 198
Trypanosomes, 189; American, 202, **203;** Chagas disease, **203;** control, 199; fish, 79; frog, 79, 80; history, 78; life cycle, 191, 194; man, 194, 197; rat, 191; reservoir hosts, 200, 204; species, 194; transmission, 81, 193; treatment, 201; Vampire bat, 80
Trypanosomiasis, 189; American, 202; California, 202; history, 196
Tryparsamide, 202
Tsetse fly, 81, **199**
Tuira river, 102
Turkey flagellates, **77**
Twist disease, 69

Uganda, 198
Undulating membrane, 132, **189,** 190
Uroglena, 4, **5**
Uta, 210
Utricularia, 111

Vaginal trichomonads, 156
Valentin, 79
Vampire bat, 80
Vedder, 169
Vermicule, **217,** 223
Vianna, 207, 208
Virchow, 242
Viscerotome, 211
Volvox, 43, **44**
Vorticella, **48,** 51
Vorticella alba, **23,** 24
Vorticellidae, 25

Wagner-Juaregg, 263
Wechselfieber, 239
Werlhof, 240
Whip-bearers, 41
White ants, **58**
Winterbottom's sign, **195,** 196
Wormy halibut, 69
Wright, 207

Yatren, 169
Yavisa, 103
Yellow fever, 255

Ziemann, 197
Zoochlorella, 57
Zoology, 136
Zoomastigina, 46
Zooxanthella, 58

SOME DOVER SCIENCE BOOKS

CHANCE, LUCK AND STATISTICS: THE SCIENCE OF CHANCE,
Horace C. Levinson

Theory of probability and science of statistics in simple, non-technical language. Part I deals with theory of probability, covering odd superstitions in regard to "luck," the meaning of betting odds, the law of mathematical expectation, gambling, and applications in poker, roulette, lotteries, dice, bridge, and other games of chance. Part II discusses the misuse of statistics, the concept of statistical probabilities, normal and skew frequency distributions, and statistics applied to various fields—birth rates, stock speculation, insurance rates, advertising, etc. "Presented in an easy humorous style which I consider the best kind of expository writing," Prof. A. C. Cohen, Industry Quality Control. Enlarged revised edition. Formerly titled *The Science of Chance.* Preface and two new appendices by the author. Index. xiv + 365pp. 5⅜ x 8. Paperbound $2.00

BASIC ELECTRONICS,
prepared by the U.S. Navy Training Publications Center

A thorough and comprehensive manual on the fundamentals of electronics. Written clearly, it is equally useful for self-study or course work for those with a knowledge of the principles of basic electricity. Partial contents: Operating Principles of the Electron Tube; Introduction to Transistors; Power Supplies for Electronic Equipment; Tuned Circuits; Electron-Tube Amplifiers; Audio Power Amplifiers; Oscillators; Transmitters; Transmission Lines; Antennas and Propagation; Introduction to Computers; and related topics. Appendix. Index. Hundreds of illustrations and diagrams. vi + 471pp. 6½ x 9¼.
Paperbound $2.75

BASIC THEORY AND APPLICATION OF TRANSISTORS,
prepared by the U.S. Department of the Army

An introductory manual prepared for an army training program. One of the finest available surveys of theory and application of transistor design and operation. Minimal knowledge of physics and theory of electron tubes required. Suitable for textbook use, course supplement, or home study. Chapters: Introduction; fundamental theory of transistors; transistor amplifier fundamentals; parameters, equivalent circuits, and characteristic curves; bias stabilization; transistor analysis and comparison using characteristic curves and charts; audio amplifiers; tuned amplifiers; wide-band amplifiers; oscillators; pulse and switching circuits; modulation, mixing, and demodulation; and additional semiconductor devices. Unabridged, corrected edition. 240 schematic drawings, photographs, wiring diagrams, etc. 2 Appendices. Glossary. Index. 263pp. 6½ x 9¼. Paperbound $1.25

GUIDE TO THE LITERATURE OF MATHEMATICS AND PHYSICS,
N. G. Parke III

Over 5000 entries included under approximately 120 major subject headings of selected most important books, monographs, periodicals, articles in English, plus important works in German, French, Italian, Spanish, Russian (many recently available works). Covers every branch of physics, math, related engineering. Includes author, title, edition, publisher, place, date, number of volumes, number of pages. A 40-page introduction on the basic problems of research and study provides useful information on the organization and use of libraries, the psychology of learning, etc. This reference work will save you hours of time. 2nd revised edition. Indices of authors, subjects, 464pp. 5⅜ x 8.
Paperbound $2.75

THE RISE OF THE NEW PHYSICS (formerly THE DECLINE OF MECHANISM), *A. d'Abro*
This authoritative and comprehensive 2-volume exposition is unique in scientific publishing. Written for intelligent readers not familiar with higher mathematics, it is the only thorough explanation in non-technical language of modern mathematical-physical theory. Combining both history and exposition, it ranges from classical Newtonian concepts up through the electronic theories of Dirac and Heisenberg, the statistical mechanics of Fermi, and Einstein's relativity theories. "A must for anyone doing serious study in the physical sciences," *J. of Franklin Inst.* 97 illustrations. 991pp. 2 volumes.
Vol. 1 Paperbound $2.25, Vol. 2 Paperbound $2.25,
The set $4.50

THE STRANGE STORY OF THE QUANTUM, AN ACCOUNT FOR THE GENERAL READER OF THE GROWTH OF IDEAS UNDERLYING OUR PRESENT ATOMIC KNOWLEDGE, *B. Hoffmann*
Presents lucidly and expertly, with barest amount of mathematics, the problems and theories which led to modern quantum physics. Dr. Hoffmann begins with the closing years of the 19th century, when certain trifling discrepancies were noticed, and with illuminating analogies and examples takes you through the brilliant concepts of Planck, Einstein, Pauli, de Broglie, Bohr, Schroedinger, Heisenberg, Dirac, Sommerfeld, Feynman, etc. This edition includes a new, long postscript carrying the story through 1958. "Of the books attempting an account of the history and contents of our modern atomic physics which have come to my attention, this is the best," H. Margenau, Yale University, in *American Journal of Physics.* 32 tables and line illustrations. Index. 275pp. 5⅜ x 8.
Paperbound $1.75

GREAT IDEAS AND THEORIES OF MODERN COSMOLOGY, *Jagjit Singh*
The theories of Jeans, Eddington, Milne, Kant, Bondi, Gold, Newton, Einstein, Gamow, Hoyle, Dirac, Kuiper, Hubble, Weizsäcker and many others on such cosmological questions as the origin of the universe, space and time, planet formation, "continuous creation," the birth, life, and death of the stars, the origin of the galaxies, etc. By the author of the popular *Great Ideas of Modern Mathematics.* A gifted popularizer of science, he makes the most difficult abstractions crystal-clear even to the most non-mathematical reader. Index. xii + 276pp. 5⅜ x 8½. Paperbound $2.00

GREAT IDEAS OF MODERN MATHEMATICS: THEIR NATURE AND USE, *Jagjit Singh*
Reader with only high school math will understand main mathematical ideas of modern physics, astronomy, genetics, psychology, evolution, etc., better than many who use them as tools, but comprehend little of their basic structure. Author uses his wide knowledge of non-mathematical fields in brilliant exposition of differential equations, matrices, group theory, logic, statistics, problems of mathematical foundations, imaginary numbers, vectors, etc. Original publications, 2 appendices. 2 indexes. 65 illustr. 322pp. 5⅜ x 8. Paperbound $2.00

THE MATHEMATICS OF GREAT AMATEURS, *Julian L. Coolidge*
Great discoveries made by poets, theologians, philosophers, artists and other non-mathematicians: Omar Khayyam, Leonardo da Vinci, Albrecht Dürer, John Napier, Pascal, Diderot, Bolzano, etc. Surprising accounts of what can result from a non-professional preoccupation with the oldest of sciences. 56 figures. viii + 211pp. 5⅜ x 8½. Paperbound $1.50

DIFFERENTIAL EQUATIONS,
F. R. Moulton
A detailed, rigorous exposition of all the non-elementary processes of solving ordinary differential equations. Several chapters devoted to the treatment of practical problems, especially those of a physical nature, which are far more advanced than problems usually given as illustrations. Includes analytic differential equations; variations of a parameter; integrals of differential equations; analytic implicit functions; problems of elliptic motion; sine-amplitude functions; deviation of formal bodies; Cauchy-Lipschitz process; linear differential equations with periodic coefficients; differential equations in infinitely many variations; much more. Historical notes. 10 figures. 222 problems. Index. xv + 395pp. 5⅜ x 8. Paperbound $2.00

PARTIAL DIFFERENTIAL EQUATIONS OF MATHEMATICAL PHYSICS,
A. G. Webster
A keystone work in the library of every mature physicist, engineer, researcher. Valuable sections on elasticity, compression theory, potential theory, theory of sound, heat conduction, wave propagation, vibration theory. Contents include: deduction of differential equations, vibrations, normal functions, Fourier's series, Cauchy's method, boundary problems, method of Riemann-Volterra, spherical, cylindrical, ellipsoidal harmonics, applications, etc. 97 figures. vii + 440pp. 5⅜ x 8. Paperbound $2.25

THE CONTINUUM AND OTHER TYPES OF SERIAL ORDER,
E. V. Huntington
This famous book gives a systematic elementary account of the modern theory of the continuum as a type of serial order. Based on the Cantor-Dedekind ordinal theory, which requires no technical knowledge of higher mathematics, it offers an easily followed analysis of ordered classes, discrete and dense series, continuous series, Cantor's transfinite numbers. 2nd edition. Index. viii + 82pp. 5⅜ v 8. Paperbound $1.00

CONTRIBUTIONS TO THE FOUNDING OF THE THEORY OF TRANSFINITE NUMBERS, *Georg Cantor*
These papers founded a new branch of mathematics. The famous articles of 1895-7 are translated, with an 82-page introduction by P. E. B. Jourdain dealing with Cantor, the background of his discoveries, their results, future possibilities. Bibliography. Index. Notes. ix + 211pp. 5⅜ x 8. Paperbound $1.35

ADVANCED EUCLIDEAN GEOMETRY,
R. A. Johnson
For years the standard textbook on advanced Euclidean geometry, requires only high school geometry and trigonometry. Explores in unusual detail and gives proofs of hundreds of relatively recent theorems and corollaries, many formerly available only in widely scattered journals. Covers tangent circles, the theorem of Miquel, symmedian point, pedal triangles and circles, the Brocard configuration, and much more. Formerly "Modern Geometry." Index. 107 diagrams. xiii + 319pp. 5⅜ x 8. Paperbound $1.65

Prices subject to change without notice.

Available at your book dealer or write for free catalogue to Dept. Adsci, Dover Publications, Inc., 180 Varick St., N.Y., N.Y. 10014. Dover publishes more than 150 books each year on science, elementary and advanced mathematics, biology, music, art, literary history, social sciences and other areas.